FREE TO CHOOSE

Other Books by Milton Friedman

MONEY MISCHIEF:
EPISODES IN MONETARY HISTORY

PRICE THEORY

THERE IS NO SUCH THING AS A FREE LUNCH

AN ECONOMIST'S PROTEST

THE OPTIMUM QUANTITY OF MONEY
AND OTHER ESSAYS

DOLLARS AND DEFICITS

A MONETARY HISTORY OF THE UNITED STATES
(with Anna J. Schwartz)

INFLATION: CAUSES AND CONSEQUENCES

BRIGHT PROMISES, DISMAL PERFORMANCE*

With Rose Friedman

CAPITALISM AND FREEDOM

TYRANNY OF THE STATUS QUO

**Also available from Harcourt Brace & Company
in a Harvest paperback edition*

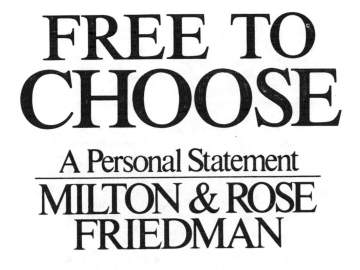

FREE TO CHOOSE

A Personal Statement

MILTON & ROSE FRIEDMAN

A HARVEST BOOK
HARCOURT BRACE & COMPANY
San Diego   New York   London

The authors wish to thank the following publishers for permission to quote from the sources listed:

Harvard Educational Review. Excerpts from "Alternative Public School Systems" by Kenneth B. Clark in the *Harvard Educational Review,* Winter 1968. Reprinted by permission of the publisher.
Newsweek magazine. Excerpt from "Barking Cats" by Milton Friedman in *Newsweek* magazine, February 19, 1973. Copyright © 1973 by Newsweek, Inc. All rights reserved. Reprinted by permission.
The Wall Street Journal. Excerpts from "The Swedish Tax Revolt" by Melvyn B. Krauss in *The Wall Street Journal,* February 1, 1979. Reprinted by permission of The Wall Street Journal, © Dow Jones & Co., Inc., 1979. All rights reserved.

Library of Congress Cataloging-in-Publication Data
Friedman, Milton, 1912–
Free to choose: a personal statement/Milton & Rose Friedman.
p.   cm.
Includes bibliographical references and index.
ISBN 0-15-633460-7 (pbk.)
1. Capitalism. 2. Welfare state. 3. Industry and state.
I. Friedman, Rose D. II. Title.
[HB501.F72   1990]   330.12'2—dc20   90-36179

Printed in the United States of America

First Harvest edition 1990
G I J H F

To Ricky and Patri

CONTENTS

FOREWORD TO THE HARVEST EDITION

When *Free to Choose* was first published a decade ago, we were sufficiently optimistic to label our final chapter "The Tide Is Turning." The climate of opinion was, we thought, shifting away from a belief in collectivism and toward a belief in individualism and private markets. We did not dream that the tide would turn as dramatically as it has—on both sides of the Iron Curtain.

Ten years ago, many people around the world believed that socialism was a viable, even the most promising, system for promoting material prosperity and human freedom. Few people anywhere in the world believe that today. Idealistic faith in socialism still lives on, but only in some ivory tower enclaves in the West and in some of the most backward countries elsewhere. Ten years ago, many people were convinced that capitalism, based on free private markets, was a deeply flawed system that was not capable of achieving both widely shared prosperity and human freedom. Today conventional wisdom regards capitalism as the only system that can do so.

Is *Free to Choose* outdated and no longer needed now that its main thesis has become conventional wisdom? Far from it. Conventional wisdom may have changed, but conventional practice has not. Political leaders in capitalist countries who cheer the collapse of socialism in other countries continue to favor socialist solutions in their own. They know the words, but they have not learned the tune.

Despite the drastic change in intellectual and popular opinion in the past decade, governments of so-called capitalist countries are just as backward as governments of communist countries in dismantling the socialist practices that have mushroomed in recent decades. The fraction of our income that goes to finance government spending—supposedly on our behalf—has not declined appreciably and in many countries has continued to increase. In the United States, it was 40 percent in 1980 and 42 percent in 1988, down from a high of 44 percent in 1986. Neither has there been much letup in the flood of detailed regulations that control our lives: in 1980, 87,012 pages were added to the *Federal Register*, which records all regulations; in 1988, 53,376 pages. In the words of the Declaration of Independence, our governments continue to erect "a multitude of new offices" and send "swarms of officers to harass our people and eat out our substance."

Restraints on international trade, analyzed in Chapter 2 of *Free to Choose*, have grown, not declined; some restrictions on prices and wages, particularly exchange controls, have been eliminated or reduced, but others have been added. Our cradle-to-grave Social Security system has become more extensive and is in greater need of reform than ever (Chapter 4); that is equally true of our school system (Chapter 6). The institutions set up to "protect the consumer" and "the worker" continue to have effects opposite to those intended by their well-meaning sponsors (Chapters 7 and 8). In these and other areas, the momentum of past practices has overwhelmed the effect of a changed climate of opinion.

There has been a substantial improvement in the rate of inflation, which has declined worldwide—in the U.S., from well over 10 percent a year to less than 5 percent. However, inflation is by no means conquered, and our analysis of the causes, consequences, and cure of inflation in Chapter 9 remains valid and highly relevant to assure that recent reductions in inflation are more than a flash in the pan.

The big change has been not in achievements but in prospects. Free private markets are far more likely to multiply in coming years than seemed possible ten years ago. As a result, a book that explains how free private markets work, what their advantages are, and how to eliminate obstacles to their more effective operation has even greater relevance today than it had ten years ago.

Some specific figures and references in our book are now out of date, but we have thought it best to reprint the manuscript essentially unchanged. A thorough revision of the book to bring it up to date and to include the new problems that have arisen in the interim might well be worthwhile, but we have not been in a position to undertake the task and concluded that it was better to leave the manuscript alone than to do a superficial update. We hope that the occasional anachronism will not interfere with the reader's understanding.

What appeared to many readers of the book ten years ago as utopian and unrealistic will, we believe, appear to many new readers as almost a blueprint for practical change. We are therefore delighted that Harcourt Brace Jovanovich is issuing a new edition of *Free to Choose*. The tide has turned, but it is still far from the flood tide that is so badly needed to assure a bright future for human freedom.

Milton and Rose Friedman
January 4, 1990

PREFACE

This book has two parents: *Capitalism and Freedom,* our earlier book, published in 1962 (University of Chicago Press); and a TV series, titled, like the book, "Free to Choose." The series will be shown on the Public Broadcasting Service for ten successive weeks in 1980.

Capitalism and Freedom examines "the role of competitive capitalism—the organization of the bulk of economic activity through private enterprise operating in a free market—as a system of economic freedom and a necessary condition for political freedom." In the process, it defines the role that government should play in a free society.

"Our principles offer," *Capitalism and Freedom* says, "no hard and fast line how far it is appropriate to use government to accomplish jointly what it is difficult or impossible for us to accomplish separately through strictly voluntary exchange. In any particular case of proposed intervention, we must make up a balance sheet, listing separately the advantages and disadvantages. Our principles tell us what items to put on the one side and what items on the other and they give us some basis for attaching importance to the different items."

To give substance to those principles and illustrate their application, *Capitalism and Freedom* examines specific issues—among others, monetary and fiscal policy, the role of government in education, capitalism and discrimination, and the alleviation of poverty.

Free to Choose is a less abstract and more concrete book. Readers of *Capitalism and Freedom* will find here a fuller development of the philosophy that permeates both books—here, there are more nuts and bolts, less theoretical framework. Moreover, this book is influenced by a fresh approach to political science that has come mainly from economists—Anthony Downs,

James M. Buchanan, Gordon Tullock, George J. Stigler, and Gary S. Becker, who, along with many others, have been doing exciting work in the economic analysis of politics. *Free to Choose* treats the political system symmetrically with the economic system. Both are regarded as markets in which the outcome is determined by the interaction among persons pursuing their own self-interests (broadly interpreted) rather than by the social goals the participants find it advantageous to enunciate. That is implicit throughout the book and explicit in the final chapter.

The TV series covers the same topics as this book: the ten chapters of the book correspond to the ten programs of the TV series and (except for the final chapter) bear the same titles. However, the TV series and the book are very different—each true to its own character. The book covers many items that the time constraints of the TV programs made it necessary to omit or allude to only briefly. And its coverage is more systematic and thorough.

We were induced to undertake the TV series in early 1977 by Robert Chitester, president of PBS station WQLN of Erie, Pennsylvania. His imagination and hard work, and his commitment to the values of a free society, made the series possible. At his suggestion, Milton presented between September of 1977 and May of 1978 fifteen public lectures before various audiences followed by question-and-answer sessions, all of which were videotaped. William Jovanovich committed Harcourt Brace Jovanovich to the marketing of the videotapes and provided a generous advance to help finance the videotaping of the lectures, which are currently being distributed by Harcourt Brace Jovanovich, Inc. The transcripts of the lectures served as raw material for designing the TV programs themselves.

Before the lectures were completed, Bob Chitester had succeeded in obtaining sufficient financial support to permit us to proceed with the TV series. We selected Video-Arts of London as the best group to produce it. After months of preliminary planning, actual filming began in March of 1978 and was not completed until September of 1979.

Anthony Jay, Michael Peacock, and Robert Reid of Video-Arts played a key role in the initial design of the series and an important supervisory role thereafter.

Five TV professionals were with us throughout most of the filming and editing: Michael Latham, as producer of the series; Graham Massey, as film director; Eben Wilson, as an associate producer and principal researcher; Margaret Young, as assistant film director and production secretary; and Jackie Warner, as production manager. They initiated us gently but firmly into the arcane art of making TV documentaries and smoothed over the difficult spots with invariable tact and friendship. They made our venture into a strange and complex world an exciting and enjoyable experience rather than the nightmare that we now realize it could easily have become.

Their insistence on combining brevity with both rigor and lucidity forced us to rethink many of our own ideas and to pare them down to essentials. The discussions with them, as well as with the film crews from different countries—one of the most enjoyable parts of the project—helped us to recognize weak points in our reasoning and induced us to search for further evidence. Released from the rigid time constraints of TV, we have been able to take full advantage of these discussions in this book.

We are in debt to Edward C. Banfield and David D. Friedman, who read the complete first draft, and to George Stigler, Aaron Director, Chiaki Nishiyama, Colin Campbell, and Anna Schwartz. Rosemary Campbell spent many hours of painstaking work in the library checking facts and figures. We cannot blame her if errors do appear, for we did some of the checking ourselves. We owe much to Gloria Valentine, Milton's secretary, whose good nature is matched by her competence. Finally, we appreciate the help we have received from Harcourt Brace Jovanovich, some anonymously, some from William Jovanovich, Carol Hill, and our editor, Peggy Brooks.

Television is dramatic. It appeals to the emotions. It captures your attention. Yet, we remain of the opinion that the printed page is a more effective instrument for both education and persuasion. The authors of a book can explore issues deeply—without being limited by the ticking clock. The reader can stop and think, turn the pages back without being diverted by the emotional appeal of the scenes moving relentlessly across his television screen.

Anyone who is persuaded in one evening (or even ten one-hour evenings) is not really persuaded. He can be converted by the next person of opposite views with whom he spends an evening. The only person who can truly persuade you is yourself. You must turn the issues over in your mind at leisure, consider the many arguments, let them simmer, and after a long time turn your preferences into convictions.

Milton Friedman
Rose D. Friedman
Ely, Vermont
September 28, 1979

"Experience should teach us to be most on our guard to protect liberty when the government's purposes are beneficial. Men born to freedom are naturally alert to repel invasion of their liberty by evil-minded rulers. The greater dangers to liberty lurk in insidious encroachment by men of zeal, well-meaning but without understanding."

—Justice Louis Brandeis,
Olmstead v. *United States*,
277 U.S. 479 (1928)

INTRODUCTION

Ever since the first settlement of Europeans in the New World America has been a magnet for people seeking adventure, fleeing from tyranny, or simply trying to make a better life for themselves and their children.

An initial trickle swelled after the American Revolution and the establishment of the United States of America and became a flood in the nineteenth century, when millions of people streamed across the Atlantic, and a smaller number across the Pacific, driven by misery and tyranny, and attracted by the promise of freedom and affluence.

When they arrived, they did not find streets paved with gold; they did not find an easy life. They did find freedom and an opportunity to make the most of their talents. Through hard work, ingenuity, thrift, and luck, most of them succeeded in realizing enough of their hopes and dreams to encourage friends and relatives to join them.

The story of the United States is the story of an economic miracle and a political miracle that was made possible by the translation into practice of two sets of ideas—both, by a curious coincidence, formulated in documents published in the same year, 1776.

One set of ideas was embodied in *The Wealth of Nations*, the masterpiece that established the Scotsman Adam Smith as the father of modern economics. It analyzed the way in which a market system could combine the freedom of individuals to pursue their own objectives with the extensive cooperation and collaboration needed in the economic field to produce our food, our clothing, our housing. Adam Smith's key insight was that both parties to an exchange can benefit and that, *so long as cooperation is strictly voluntary*, no exchange will take place unless both parties do benefit. No external force, no coercion, no violation of freedom is necessary to produce cooperation among individuals all of whom can benefit. That is why, as Adam Smith put it, an individual who "intends only his own gain" is "led by an invisible hand to promote an end which was no part of his intention. Nor is it always the worse for the society that it was no part of it. By pursuing his own interest he frequently promotes that of the society more effectually than when he really intends to promote it. I have never known much good done by those who affected to trade for the public good." [1]

The second set of ideas was embodied in the Declaration of Independence, drafted by Thomas Jefferson to express the general sense of his fellow countrymen. It proclaimed a new nation, the first in history established on the principle that every person is entitled to pursue his own values: "We hold these truths to be self-evident, that all men are created equal, that they are endowed by their Creator with certain unalienable Rights; that among these are Life, Liberty, and the pursuit of Happiness."

Or, as stated in more extreme and unqualified form nearly a century later by John Stuart Mill,

> The sole end for which mankind are warranted, individually or collectively, in interfering with the liberty of action of any of their number, is self protection. . . . [T]he only purpose for which power can be rightfully exercised over any member of a civilized community, against his will, is to prevent harm to others. His own good, either physical or moral, is not a sufficient warrant. . . . The only part of the conduct of any one, for which he is amenable to society, is that which concerns others. In the part which merely concerns himself, his independence is, of right, absolute. Over himself, over his own body and mind, the individual is sovereign.[2]

Much of the history of the United States revolves about the attempt to translate the principles of the Declaration of Independence into practice—from the struggle over slavery, finally settled by a bloody civil war, to the subsequent attempt to promote equality of opportunity, to the more recent attempt to achieve equality of results.

Economic freedom is an essential requisite for political freedom. By enabling people to cooperate with one another without coercion or central direction, it reduces the area over which political power is exercised. In addition, by dispersing power, the free market provides an offset to whatever concentration of political power may arise. The combination of economic and political *power* in the same hands is a sure recipe for tyranny.

The combination of economic and political *freedom* produced a golden age in both Great Britain and the United States in the nineteenth century. The United States prospered even more than Britain. It started with a clean slate: fewer vestiges of class and status; fewer government restraints; a more fertile field for energy, drive, and innovation; and an empty continent to conquer.

The fecundity of freedom is demonstrated most dramatically and clearly in agriculture. When the Declaration of Independence was enacted, fewer than 3 million persons of European and African origin (i.e., omitting the native Indians) occupied a narrow fringe along the eastern coast. Agriculture was the main economic activity. It took nineteen out of twenty workers to feed the country's inhabitants and provide a surplus for export in exchange for foreign goods. Today it takes fewer than one out of twenty workers to feed the 220 million inhabitants and provide a surplus that makes the United States the largest single exporter of food in the world.

What produced this miracle? Clearly not central direction by government—nations like Russia and its satellites, mainland China, Yugoslavia, and India that today rely on central direction employ from one-quarter to one-half of their workers in agriculture, yet frequently rely on U.S. agriculture to avoid mass starvation. During most of the period of rapid agricultural expansion in the United States the government played a negligible role. Land was made available—but it was land that had been unproductive before. After the middle of the nineteenth century land-grant colleges were established, and they disseminated information and technology through governmentally financed extension services. Unquestionably, however, the main source of the agricultural revolution was private initiative operating in a free market open to all—the shame of slavery only excepted. And the most rapid growth came after slavery was abolished. The millions of immigrants from all over the world were free to work for themselves, as independent farmers or businessmen, or to work for others, at terms mutually agreed. They were free to experiment with new techniques—at their risk if the experiment failed, and to their profit if it succeeded. They got little assistance from government. Even more important, they encountered little interference from government.

Government started playing a major role in agriculture during and after the Great Depression of the 1930s. It acted primarily to restrict output in order to keep prices artificially high.

The growth of agricultural productivity depended on the accompanying industrial revolution that freedom stimulated. Thence came the new machines that revolutionized agriculture. Conversely, the industrial revolution depended on the availability of the manpower released by the agricultural revolution. Industry and agriculture marched hand in hand.

Smith and Jefferson alike had seen concentrated government power as a great danger to the ordinary man; they saw the protection of the citizen against the tyranny of government as the perpetual need. That was the aim of the Virginia Declaration of Rights (1776) and the United States Bill of Rights (1791); the purpose of the separation of powers in the U.S. Constitution; the moving force behind the changes in the British legal structure from the issuance of the Magna Carta in the thirteenth century to the end of the nineteenth century. To Smith and Jefferson, government's role was as an umpire, not a participant. Jefferson's ideal, as he expressed it in his first inaugural address (1801), was "[a] wise and frugal government, which shall restrain men from injuring one another, which shall leave them otherwise free to regulate their own pursuits of industry and improvement."

Ironically, the very success of economic and political freedom reduced its appeal to later thinkers. The narrowly limited government of the late nineteenth century possessed little concentrated power that endangered the ordinary man. The other side of that coin was that it possessed little power that would enable good people to do good. And in an imperfect world there were still many evils. Indeed, the very progress of society made the residual evils seem all the more objectionable. As always, people took the favorable developments for granted. They forgot the danger to freedom from a strong government. Instead, they were attracted by the good that a stronger government could achieve—if only government power were in the "right" hands.

These ideas began to influence government policy in Great Britain by the beginning of the twentieth century. They gained increasing acceptance among intellectuals in the United States but had little effect on government policy until the Great Depression of the early 1930s. As we show in Chapter 3, the depression was produced by a failure of government in one area—money —where it had exercised authority ever since the beginning of the Republic. However, government's responsibility for the depression was not recognized—either then or now. Instead, the depression was widely interpreted as a failure of free market capitalism. That myth led the public to join the intellectuals in a changed view of the relative responsibilities of individuals and government. Emphasis on the responsibility of the individual for his own fate was replaced by emphasis on the individual as a pawn buffeted by forces beyond his control. The view that government's role is to serve as an umpire to prevent individuals from coercing one another was replaced by the view that government's role is to serve as a parent charged with the duty of coercing some to aid others.

These views have dominated developments in the United States during the past half-century. They have led to a growth in government at all levels, as well as to a transfer of power from local government and local control to central government and central control. The government has increasingly undertaken the task of taking from some to give to others in the name of security and equality. One government policy after another has been set up to "regulate" our "pursuits of industry and improvement," standing Jefferson's dictum on its head (Chapter 7).

These developments have been produced by good intentions with a major assist from self-interest. Even the strongest supporters of the welfare and paternal state agree that the results have been disappointing. In the government sphere, as in the market, there seems to be an invisible hand, but it operates in precisely the opposite direction from Adam Smith's: an individual who intends only to serve the public interest by fostering government intervention is "led by an invisible hand to promote" private interests, "which was no part of his intention." That conclusion is driven home again and again as we examine, in the chapters that follow, the several areas in which government power has been exercised—whether to achieve security (Chapter 4) or equality (Chapter 5), to promote education (Chapter 6), to protect the consumer (Chapter 7) or the worker (Chapter 8), or to avoid inflation and promote employment (Chapter 9).

So far, in Adam Smith's words, "the uniform, constant, and uninterrupted effort of every man to better his condition, the principle from which public and national, as well as private opulence is originally derived," has been "powerful enough to maintain the natural progress of things toward improvement, in spite both of the extravagance of governments and of the greatest errors of administration. Like the unknown principle of animal life, it frequently restores health and vigour to the constitution, in spite, not only of the disease, but of the absurd prescriptions of the doctor." [3] So far, that is, Adam Smith's invisible hand has been powerful enough to overcome the deadening effects of the invisible hand that operates in the political sphere.

The experience of recent years—slowing growth and declining productivity—raises a doubt whether private ingenuity can continue to overcome the deadening effects of government control if we continue to grant ever more power to government, to authorize a "new class" of civil servants to spend ever larger fractions of our income supposedly on our behalf. Sooner or later—and perhaps sooner than many of us expect—an ever bigger government would destroy both the prosperity that we owe to the free market and the human freedom proclaimed so eloquently in the Declaration of Independence.

We have not yet reached the point of no return. We are still free as a people to choose whether we shall continue speeding down the "road to serfdom," as Friedrich Hayek entitled his profound and influential book, or whether we shall set tighter limits on government and rely more heavily on voluntary cooperation among free individuals to achieve our several objectives. Will our golden age come to an end in a relapse into the tyranny and misery that has always been, and remains today, the state of most of mankind? Or shall we have the wisdom, the foresight, and the courage to change our course, to learn from experience, and to benefit from a "rebirth of freedom"?

If we are to make that choice wisely, we must understand the fundamental principles of our system, both the economic principles of Adam Smith, which explain how it is that a complex, organized, smoothly running system can develop and flourish without central direction, how coordination can be achieved without coercion (Chapter 1); and the political principles expressed by Thomas Jefferson (Chapter 5). We must understand why it is that attempts to replace cooperation by central direction are capable of doing so much harm (Chapter 2). We must understand also the intimate connection between political freedom and economic freedom.

Fortunately, the tide is turning. In the United States, in Great Britain, the countries of Western Europe, and in many other countries around the world, there is growing recognition of the dangers of big government, growing dissatisfaction with the policies that have been followed. This shift is being reflected not only in opinion, but also in the political sphere. It is becoming politically profitable for our representatives to sing a different tune—and perhaps even to act differently. We are experiencing another major change in public opinion. We have the opportunity to nudge the change in opinion toward greater reliance on individual initiative and voluntary cooperation, rather than toward the other extreme of total collectivism.

In our final chapter, we explore why it is that in a supposedly democratic political system special interests prevail over the general interest. We explore what we can do to correct the defect in our system that accounts for that result, how we can limit government while enabling it to perform its essential functions of defending the nation from foreign enemies, protecting each of us from coercion by our fellow citizens, adjudicating our disputes, and enabling us to agree on the rules that we shall follow.

CHAPTER 1

The Power
of the Market

Every day each of us uses innumerable goods and services—to eat, to wear, to shelter us from the elements, or simply to enjoy. We take it for granted that they will be available when we want to buy them. We never stop to think how many people have played a part in one way or another in providing those goods and services. We never ask ourselves how it is that the corner grocery store—or nowadays, supermarket—has the items on its shelves that we want to buy, how it is that most of us are able to earn the money to buy those goods.

It is natural to assume that someone must give orders to make sure that the "right" products are produced in the "right" amounts and available at the "right" places. That is one method of coordinating the activities of a large number of people—the method of the army. The general gives orders to the colonel, the colonel to the major, the major to the lieutenant, the lieutenant to the sergeant, and the sergeant to the private.

But that command method can be the exclusive or even principal method of organization only in a very small group. Not even the most autocratic head of a family can control every act of other family members entirely by order. No sizable army can really be run entirely by command. The general cannot conceivably have the information necessary to direct every movement of the lowliest private. At every step in the chain of command, the soldier, whether officer or private, must have discretion to take into account information about specific circumstances that his commanding officer could not have. Commands must be supplemented by voluntary cooperation—a less obvious and more subtle, but far more fundamental, technique of coordinating the activities of large numbers of people.

Russia is the standard example of a large economy that is supposed to be organized by command—a centrally planned economy. But that is more fiction than fact. At every level of the economy, voluntary cooperation enters to supplement central planning or to offset its rigidities—sometimes legally, sometimes illegally.[1]

In agriculture, full-time workers on government farms are permitted to grow food and raise animals on small private plots in their spare time for their own use or to sell in relatively free markets. These plots account for less than 1 percent of the agricultural land in the country, yet they are said to provide nearly a third of total farm output in the Soviet Union (are "said to" because it is likely that some products of government farms are clandestinely marketed as if from private plots).

In the labor market individuals are seldom ordered to work at specific jobs; there is little actual direction of labor in this sense. Rather, wages are offered for various jobs, and individuals apply for them—much as in capitalist countries. Once hired, they may subsequently be fired or may leave for jobs they prefer. Numerous restrictions affect who may work where, and, of course, the laws prohibit anyone from setting up as an employer—although numerous clandestine workshops serve the extensive black market. Allocation of workers on a large scale primarily by compulsion is just not feasible; and neither, apparently, is complete suppression of private entrepreneurial activity.

The attractiveness of different jobs in the Soviet Union often depends on the opportunities they offer for extralegal or illegal moonlighting. A resident of Moscow whose household equipment fails may have to wait months to have it repaired if he calls the state repair office. Instead, he may hire a moonlighter—very likely someone who works for the state repair office. The householder gets his equipment repaired promptly; the moonlighter gets some extra income. Both are happy.

These voluntary market elements flourish despite their inconsistency with official Marxist ideology because the cost of eliminating them would be too high. Private plots could be forbidden—but the famines of the 1930s are a stark reminder of the cost. The Soviet economy is hardly a model of efficiency now. Without the voluntary elements it would operate at an even lower level of effectiveness. Recent experience in Cambodia tragically illustrates the cost of trying to do without the market entirely.

Just as no society operates entirely on the command principle, so none operates entirely through voluntary cooperation. Every society has some command elements. These take many forms. They may be as straightforward as military conscription or forbidding the purchase and sale of heroin or cyclamates or court orders to named defendants to desist from or perform specified actions. Or, at the other extreme, they may be as subtle as imposing a heavy tax on cigarettes to discourage smoking—a hint, if not a command, by some of us to others of us.

It makes a vast difference what the mix is—whether voluntary exchange is primarily a clandestine activity that flourishes because of the rigidities of a dominant command element, or whether voluntary exchange is the dominant principle of organization, supplemented to a smaller or larger extent by command elements. Clandestine voluntary exchange may prevent a command economy from collapsing, may enable it to creak along and even achieve some progress. It can do little to undermine the tyranny on which a predominantly command economy rests. A predominantly voluntary exchange economy, on the other hand, has within it the potential to promote both prosperity and human freedom. It may not achieve its potential in either respect, but we know of no society that has ever achieved prosperity and freedom unless voluntary exchange has been its dominant principle of organization. We hasten to add that voluntary exchange is not a sufficient condition for prosperity and freedom. That, at least, is the lesson of history to date. Many societies organized predominantly by voluntary exchange have not achieved either prosperity or freedom, though they have achieved a far greater measure of both than authoritarian societies. But voluntary exchange is a necessary condition for both prosperity and freedom.

COOPERATION THROUGH VOLUNTARY EXCHANGE

A delightful story called "I, Pencil: My Family Tree as Told to Leonard E. Read" [2] dramatizes vividly how voluntary exchange enables millions of people to cooperate with one another. Mr. Read, in the voice of the "Lead Pencil—the ordinary wooden pencil familiar to all boys and girls and adults who can read and write," starts his story with the fantastic statement that "*not a single person . . . knows how to make me."* Then he proceeds to tell about all the things that go into the making of a pencil. First, the wood comes from a tree, "a cedar of straight grain that grows in Northern California and Oregon." To cut down the tree and cart the logs to the railroad siding requires "saws and trucks and rope and . . . countless other gear." Many persons and numberless skills are involved in their fabrication: in "the mining of ore, the making of steel and its refinement into saws, axes, motors; the growing of hemp and bringing it through all the stages to heavy and strong rope; the logging camps with their beds and mess halls, . . . untold thousands of persons had a hand in every cup of coffee the loggers drink!"

And so Mr. Read goes on to the bringing of the logs to the mill, the millwork involved in converting the logs to slats, and the transportation of the slats from California to Wilkes-Barre, where the particular pencil that tells the story was manufactured. And so far we have only the outside wood of the pencil. The "lead" center is not really lead at all. It starts as graphite mined in Ceylon. After many complicated processes it ends up as the lead in the center of the pencil.

The bit of metal—the ferrule—near the top of the pencil is brass. "Think of all the persons," he says, "who mine zinc and copper and those who have the skills to make shiny sheet brass from these products of nature."

What we call the eraser is known in the trade as "the plug." It is thought to be rubber. But Mr. Read tells us the rubber is only for binding purposes. The erasing is actually done by "Factice," a rubberlike product made by reacting rape seed oil from the Dutch East Indies (now Indonesia) with sulfur chloride.

After all of this, says the pencil, "Does anyone wish to challenge my earlier assertion that no single person on the face of this earth knows how to make me?"

None of the thousands of persons involved in producing the pencil performed his task because he wanted a pencil. Some among them never saw a pencil and would not know what it is for. Each saw his work as a way to get the goods and services he wanted— goods and services we produced in order to get the pencil we wanted. Every time we go to the store and buy a pencil, we are exchanging a little bit of our services for the infinitesimal amount of services that each of the thousands contributed toward producing the pencil.

It is even more astounding that the pencil was ever produced. No one sitting in a central office gave orders to these thousands of people. No military police enforced the orders that were not given. These people live in many lands, speak different languages, practice different religions, may even hate one another—yet none of these differences prevented them from cooperating to produce a pencil. How did it happen? Adam Smith gave us the answer two hundred years ago.

THE ROLE OF PRICES

The key insight of Adam Smith's *Wealth of Nations* is misleadingly simple: if an exchange between two parties is voluntary, it will not take place unless both believe they will benefit from it. Most economic fallacies derive from the neglect of this simple insight, from the tendency to assume that there is a fixed pie, that one party can gain only at the expense of another.

This key insight is obvious for a simple exchange between two individuals. It is far more difficult to understand how it can enable people living all over the world to cooperate to promote their separate interests.

The price system is the mechanism that performs this task without central direction, without requiring people to speak to one another or to like one another. When you buy your pencil or your daily bread, you don't know whether the pencil was made or the wheat was grown by a white man or a black man, by a Chinese or an Indian. As a result, the price system enables people to cooperate peacefully in one phase of their life while each one goes about his own business in respect of everything else.

Adam Smith's flash of genius was his recognition that the prices that emerged from voluntary transactions between buyers and sellers—for short, in a free market—could coordinate the activity of millions of people, each seeking his own interest, in such a way as to make everyone better off. It was a startling idea then, and it remains one today, that economic order can emerge as the unintended consequence of the actions of many people, each seeking his own interest.

The price system works so well, so efficiently, that we are not aware of it most of the time. We never realize how well it functions until it is prevented from functioning, and even then we seldom recognize the source of the trouble.

The long gasoline lines that suddenly emerged in 1974 after the OPEC oil embargo, and again in the spring and summer of 1979 after the revolution in Iran, are a striking recent example. On both occasions there was a sharp disturbance in the supply of crude oil from abroad. But that did not lead to gasoline lines in Germany or Japan, which are wholly dependent on imported oil. It led to long gasoline lines in the United States, even though we produce much of our own oil, for one reason and one reason only: because legislation, administered by a government agency, did not permit the price system to function. Prices in some areas were kept by command below the level that would have equated the amount of gasoline available at the gas stations to the amount consumers wanted to buy at that price. Supplies were allocated to different areas of the country by command, rather than in response to the pressures of demand as reflected in price. The result was surpluses in some areas and shortages plus long gasoline lines in others. The smooth operation of the price system—which for many decades had assured every consumer that he could buy gasoline at any of a large number of service stations at his convenience and with a minimal wait—was replaced by bureaucratic improvisation.

Prices perform three functions in organizing economic activity: first, they transmit information; second, they provide an incentive to adopt those methods of production that are least costly and thereby use available resources for the most highly valued purposes; third, they determine who gets how much of the product—the distribution of income. These three functions are closely interrelated.

Transmission of Information

Suppose that, for whatever reason, there is an increased demand for lead pencils—perhaps because a baby boom increases school enrollment. Retail stores will find that they are selling more pencils. They will order more pencils from their wholesalers. The wholesalers will order more pencils from the manufacturers. The manufacturers will order more wood, more brass, more graphite —all the varied products used to make a pencil. In order to induce their suppliers to produce more of these items, they will have to offer higher prices for them. The higher prices will induce the suppliers to increase their work force to be able to meet the higher demand. To get more workers they will have to offer higher wages or better working conditions. In this way ripples spread out over ever widening circles, transmitting the information to people all over the world that there is a greater demand for pencils—or, to be more precise, for some product they are engaged in producing, for reasons they may not and need not know.

The price system transmits only the important information and only to the people who need to know. The producers of wood, for example, do not have to know whether the demand for pencils has gone up because of a baby boom or because 14,000 more government forms have to be filled out in pencil. They don't even have to know that the demand for pencils has gone up. They need to know only that someone is willing to pay more for wood and that the higher price is likely to last long enough to make it worthwhile to satisfy the demand. Both items of information are provided by market prices—the first by the current price, the second by the price offered for future delivery.

A major problem in transmitting information efficiently is to make sure that everyone who can use the information gets it without clogging the "in" baskets of those who have no use for it. The price system automatically solves this problem. The people who transmit the information have an incentive to search out the people who can use it and they are in a position to do so. People who can use the information have an incentive to get it and they are in a position to do so. The pencil manufacturer is in touch with people selling the wood he uses. He is always trying to find additional suppliers who can offer him a better product or a lower price. Similarly, the producer of wood is in touch with his customers and is always trying to find new ones. On the other hand, people who are not currently engaged in these activities and are not considering them as future activities have no interest in the price of wood and will ignore it.

The transmission of information through prices is enormously facilitated these days by organized markets and by specialized communication facilities. It is a fascinating exercise to look through the price quotations published daily in, say, the *Wall Street Journal,* not to mention the numerous more specialized trade publications. These prices mirror almost instantly what is happening all over the world. There is a revolution in some remote country that is a major producer of copper, or there is a disruption of copper production for some other reason. The current price of copper will shoot up at once. To find out how long knowledgeable people expect the supplies of copper to be affected, you need merely examine the prices for future delivery on the same page.

Few readers even of the *Wall Street Journal* are interested in more than a few of the prices quoted. They can readily ignore the rest. The *Wall Street Journal* does not provide this information out of altruism or because it recognizes how important it is for the operation of the economy. Rather, it is led to provide this information by the very price system whose functioning it facilitates. It has found that it can achieve a larger or a more profitable circulation by publishing these prices—information transmitted to it by a different set of prices.

Prices not only transmit information from the ultimate buyers to retailers, wholesalers, manufacturers, and owners of resources; they also transmit information the other way. Suppose that a forest fire or strike reduces the availability of wood. The price of wood will go up. That will tell the manufacturer of pencils that it will pay him to use less wood, and it will not pay him to produce as many pencils as before unless he can sell them for a higher price. The smaller production of pencils will enable the retailer to charge a higher price, and the higher price will inform the final user that it will pay him to wear his pencil down to a shorter stub before he discards it, or shift to a mechanical pencil. Again, he doesn't need to know why the pencil has become more expensive, only that it has.

Anything that prevents prices from expressing freely the conditions of demand or supply interferes with the transmission of accurate information. Private monopoly—control over a particular commodity by one producer or a cartel of producers—is one example. That does not prevent the transmission of information through the price system, but it does distort the information transmitted. The quadrupling of the price of oil in 1973 by the oil cartel transmitted very important information. However, the information it transmitted did not reflect a sudden reduction in the supply of crude oil, or a sudden discovery of new technical knowledge about future supplies of oil, or anything else of a physical or technical character bearing on the relative availability of oil and other sources of energy. It simply transmitted the information that a group of countries had succeeded in organizing a price-fixing and market-sharing arrangement.

Price controls on oil and other forms of energy by the U.S. government in their turn prevented information about the effect of the OPEC cartel from being transmitted accurately to users of petroleum. The result both strengthened the OPEC cartel, by preventing a higher price from leading U.S. consumers to economize on the use of oil, and required the introduction of major command elements in the United States in order to allocate the scarce supply (by a Department of Energy spending in 1979 about $10 billion and employing 20,000 people).

Important as private distortions of the price system are, these days the government is the major source of interference with a free market system—through tariffs and other restraints on international trade, domestic action fixing or affecting individual prices, including wages (see Chapter 2), government regulation of specific industries (see Chapter 7), monetary and fiscal policies producing erratic inflation (see Chapter 9), and numerous other channels.

One of the major adverse effects of erratic inflation is the introduction of static, as it were, into the transmission of information through prices. If the price of wood goes up, for example, producers of wood cannot know whether that is because inflation is raising all prices or because wood is now in greater demand or lower supply relative to other products than it was before the price hike. The information that is important for the organization of production is primarily about *relative* prices—the price of one item compared with the price of another. High inflation, and particularly highly variable inflation, drowns that information in meaningless static.

Incentives

The effective transmission of accurate information is wasted unless the relevant people have an incentive to act, and act correctly, on the basis of that information. It does no good for the producer of wood to be told that the demand for wood has gone up unless he has some incentive to react to the higher price of wood by producing more wood. One of the beauties of a free price system is that the prices that bring the information also provide both an incentive to react to the information and the means to do so.

This function of prices is intimately connected with the third function—determining the distribution of income—and cannot be explained without bringing that function into the account. The producer's income—what he gets for his activities—is determined by the difference between the amount he receives from the sale of his output and the amount he spends in order to produce it. He balances the one against the other and produces an output such that producing a little more would add as much to his costs as to his receipts. A higher price shifts this margin.

In general, the more he produces, the higher the cost of producing still more. He must resort to wood in less accessible or otherwise less favorable locations; he must hire less skilled workers or pay higher wages to attract skilled workers from other pursuits. But now the higher price enables him to bear these higher costs and so provides both the incentive to increase output and the means to do so.

Prices also provide an incentive to act on information not only about the demand for output but also about the most efficient way to produce a product. Suppose one kind of wood becomes scarcer and therefore more expensive than another. The pencil manufacturer gets that information through a rise in the price of the first kind of wood. Because his income, too, is determined by the difference between sales receipts and costs, he has an incentive to economize on that kind of wood. To take a different example, whether it is less costly for loggers to use a chain saw or handsaw depends on the price of the chain saw and the handsaw, the amount of labor required with each, and the wages of different kinds of labor. The enterprise doing the logging has an incentive to acquire the relevant technical knowledge and to combine it with the information transmitted by prices in order to minimize costs.

Or take a more fanciful case that illustrates the subtlety of the price system. The rise in the price of oil engineered by the OPEC cartel in 1973 altered slightly the balance in favor of the handsaw by raising the cost of operating a chain saw. If that seems far-fetched, consider the effect on the use of diesel-powered versus gasoline-powered trucks to haul logs out of the forests and to the sawmill.

To carry this example one step further, the higher price of oil, insofar as it was permitted to occur, raised the cost of products that used more oil relative to products that used less. Consumers had an incentive to shift from the one to the other. The most obvious examples are shifts from large cars to small ones and from heating by oil to heating by coal or wood. To go much further afield to more remote effects: insofar as the relative price of wood was raised by the higher cost of producing it or by the greater demand for wood as a substitute source of energy, the resulting higher price of lead pencils gave consumers an incentive to economize on pencils! And so on in infinite variety.

We have discussed the incentive effect so far in terms of producers and consumers. But it also operates with respect to workers and owners of other productive resources. A higher demand for wood will tend to produce a higher wage for loggers. This is a signal that labor of that type is in greater demand than before. The higher wage gives workers an incentive to act on that information. Some workers who were indifferent about being loggers or doing something else may now choose to become loggers. More young people entering the labor market may become loggers. Here, too, interference by government, through minimum wages, for example, or by trade unions, through restricting entry, may distort the information transmitted or may prevent individuals from freely acting on that information (see Chapter 8).

Information about prices—whether it be wages in different activities, the rent of land, or the return to capital from different uses—is not the only information that is relevant in deciding how to use a particular resource. It may not even be the most important information, particularly about how to use one's own labor. That decision depends in addition on one's own interests and capacities—what the great economist Alfred Marshall called the whole of the advantages and disadvantages of an occupation, monetary and nonmonetary. Satisfaction in a job may compensate for low wages. On the other hand, higher wages may compensate for a disagreeable job.

Distribution of Income

The income each person gets through the market is determined, as we have seen, by the difference between his receipts from the sale of goods and services and the costs he incurs in producing those goods and services. The receipts consist predominantly of direct payments for the productive resources we own—payments for labor or the use of land or buildings or other capital. The case of the entrepreneur—like the manufacturer of pencils—is different in form but not in substance. His income, too, depends on how much of each productive resource he owns and on the price that the market sets on the services of those resources, though in his case the major productive resource he owns may be the capacity to organize an enterprise, coordinate the resources it uses, assume risks, and so on. He may also own some of the other productive resources used in the enterprise, in which case part of his income is derived from the market price for their services. Similarly, the existence of the modern corporation does not alter matters. We speak loosely of the "corporation's income" or of "business" having an income. That is figurative language. The corporation is an intermediary between its owners—the stockholders—and the resources other than the stockholders' capital, the services of which it purchases. Only people have incomes and they derive them through the market from the resources they own, whether these be in the form of corporate stock, or of bonds, or of land, or of their personal capacity.

In countries like the United States the major productive resource is personal productive capacity—what economists call "human capital." Something like three-quarters of all income generated in the United States through market transactions takes the form of the compensation of employees (wages and salaries plus supplements), and about half the rest takes the form of the income of proprietors of farms and nonfarm enterprises, which is a mixture of payment for personal services and for owned capital.

The accumulation of physical capital—of factories, mines, office buildings, shopping centers; highways, railroads, airports, cars, trucks, planes, ships; dams, refineries, power plants; houses, refrigerators, washing machines, and so on and on in endless variety—has played an essential role in economic growth. Without that accumulation the kind of economic growth that we have enjoyed could never have occurred. Without the maintenance of inherited capital the gains made by one generation would be dissipated by the next.

But the accumulation of human capital—in the form of increased knowledge and skills and improved health and longevity —has also played an essential role. And the two have reinforced one another. The physical capital enabled people to be far more productive by providing them with the tools to work with. And the capacity of people to invent new forms of physical capital, to learn how to use and get the most out of physical capital, and to organize the use of both physical and human capital on a larger and larger scale enabled the physical capital to be more productive. Both physical and human capital must be cared for and replaced. That is even more difficult and costly for human than for physical capital—a major reason why the return to human capital has risen so much more rapidly than the return to physical capital.

The amount of each kind of resource each of us owns is partly the result of chance, partly of choice by ourselves or others. Chance determines our genes and through them affects our physical and mental capacities. Chance determines the kind of family and cultural environment into which we are born and as a result our opportunities to develop our physical and mental capacity. Chance determines also other resources we may inherit from our parents or other benefactors. Chance may destroy or enhance the resources we start with. But choice also plays an important role. Our decisions about how to use our resources, whether to work hard or take it easy, to enter one occupation or another, to engage in one venture or another, to save or spend—these may determine whether we dissipate our resources or improve and add to them. Similar decisions by our parents, by other benefactors, by millions of people who may have no direct connection with us will affect our inheritance.

The price that the market sets on the services of our resources is similarly affected by a bewildering mixture of chance and choice. Frank Sinatra's voice was highly valued in twentieth-century United States. Would it have been highly valued in twentieth-century India, if he had happened to be born and to live there? Skill as a hunter and trapper had a high value in eighteenth- and nineteenth-century America, a much lower value in twentieth-century America. Skill as a baseball player brought much higher returns than skill as a basketball player in the 1920s; the reverse is true in the 1970s. These are all matters involving chance and choice—in these examples, mostly the choices made by consumers of services that determine the relative market prices of different items. But the price we receive for the services of our resources through the market also depends on our own choices—where we choose to settle, how we choose to use those resources, to whom we choose to sell their services, and so on.

In every society, however it is organized, there is always dissatisfaction with the distribution of income. All of us find it hard to understand why we should receive less than others who seem no more deserving—or why we should be receiving more than so many others whose needs seem as great and whose deserts seem no less. The farther fields always look greener—so we blame the existing system. In a command system envy and dissatisfaction are directed at the rulers. In a free market system they are directed at the market.

One result has been an attempt to separate this function of the price system—distributing income—from its other functions—transmitting information and providing incentives. Much government activity during recent decades in the United States and other countries that rely predominantly on the market has been directed at altering the distribution of income generated by the market in order to produce a different and more equal distribution of income. There is a strong current of opinion pressing for still further steps in this direction. We discuss this movement at greater length in Chapter 5.

However we might wish it otherwise, it simply is not possible to use prices to transmit information and provide an incentive to act on that information without using prices also to affect, even if not completely determine, the distribution of income. If what a person gets does not depend on the price he receives for the services of his resources, what incentive does he have to seek out information on prices or to act on the basis of that information? If Red Adair's income would be the same whether or not he performs the dangerous task of capping a runaway oil well, why should he undertake the dangerous task? He might do so once, for the excitement. But would he make it his major activity? If your income will be the same whether you work hard or not, why should you work hard? Why should you make the effort to search out the buyer who values most highly what you have to sell if you will not get any benefit from doing so? If there is no reward for accumulating capital, why should anyone postpone to a later date what he could enjoy now? Why save? How would the existing physical capital ever have been built up by the voluntary restraint of individuals? If there is no reward for maintaining capital, why should people not dissipate any capital which they have either accumulated or inherited? If prices are prevented from affecting the distribution of income, they cannot be used for other purposes. The only alternative is command. Some authority would have to decide who should produce what and how much. Some authority would have to decide who should sweep the streets and who manage the factory, who should be the policeman and who the physician.

The intimate connection among the three functions of the

price system has manifested itself in a different way in the communist countries. Their whole ideology centers on the alleged exploitation of labor under capitalism and the associated superiority of a society based on Marx's dictum: "to each according to his needs, from each according to his ability." But the inability to run a pure command economy has made it impossible for them to separate income completely from prices.

For physical resources—land, buildings, and the like—they have been able to go farthest by making them the property of the government. But even here the effect is a lack of incentive to maintain and improve the physical capital. When everybody owns something, nobody owns it, and nobody has a direct interest in maintaining or improving its condition. That is why buildings in the Soviet Union—like public housing in the United States—look decrepit within a year or two of their construction, why machines in government factories break down and are continuously in need of repair, why citizens must resort to the black market for maintaining the capital that they have for their personal use.

For human resources the communist governments have not been able to go as far as with physical resources, though they have tried to. Even they have had to permit people to own themselves to some extent and to let them make their own decisions, and have had to let prices affect and guide those decisions and determine the income received. They have, of course, distorted those prices, prevented them from being free market prices, but they have been unable to eliminate market forces.

The obvious inefficiencies that have resulted from the command system have led to much discussion by planners in socialist countries—Russia, Czechoslovakia, Hungary, China—of the possibility of making greater use of the market in organizing production. At a conference of economists from East and West, we once heard a brilliant talk by a Hungarian Marxist economist. He had rediscovered for himself Adam Smith's invisible hand—a remarkable if somewhat redundant intellectual achievement. He tried, however, to improve on it in order to use the price system to transmit information and organize production efficiently but not to distribute income. Needless to say, he failed in theory, as the communist countries have failed in practice.

A BROADER VIEW

Adam Smith's "invisible hand" is generally regarded as referring to purchases or sales of goods or services for money. But economic activity is by no means the only area of human life in which a complex and sophisticated structure arises as an unintended consequence of a large number of individuals cooperating while each pursues his own interests.

Consider, for example, language. It is a complex structure that is continually changing and developing. It has a well-defined order, yet no central body planned it. No one decided what words should be admitted into the language, what the rules of grammar should be, which words should be adjectives, which nouns. The French Academy does try to control changes in the French language, but that was a late development. It was established long after French was already a highly structured language and it mainly serves to put the seal of approval on changes over which it has no control. There have been few similar bodies for other languages.

How did language develop? In much the same way as an economic order develops through the market—out of the voluntary interaction of individuals, in this case seeking to trade ideas or information or gossip rather than goods and services with one another. One or another meaning was attributed to a word, or words were added as the need arose. Grammatical usages developed and were later codified into rules. Two parties who want to communicate with one another both benefit from coming to a common agreement about the words they use. As a wider and wider circle of people find it advantageous to communicate with one another, a common usage spreads and is codified in dictionaries. At no point is there any coercion, any central planner who has power to command, though in more recent times government school systems have played an important role in standardizing usage.

Another example is scientific knowledge. The structure of disciplines—physics, chemistry, meteorology, philosophy, humanities, sociology, economics—was not the product of a deliberate decision by anyone. Like Topsy, it "just growed." It did so because scholars found it convenient. It is not fixed, but changes as different needs develop.

Within any discipline the growth of the subject strictly parallels the economic marketplace. Scholars cooperate with one another because they find it mutually beneficial. They accept from one another's work what they find useful. They exchange their findings—by verbal communication, by circulating unpublished papers, by publishing in journals and books. Cooperation is worldwide, just as in the economic market. The esteem or approval of fellow scholars serves very much the same function that monetary reward does in the economic market. The desire to earn that esteem, to have their work accepted by their peers, leads scholars to direct their activities in scientifically efficient directions. The whole becomes greater than the sum of its parts, as one scholar builds on another's work. His work in turn becomes the basis for further development. Modern physics is as much a product of a free market in ideas as a modern automobile is a product of a free market in goods. Here again, developments have been much influenced, particularly recently, by government involvement, which has affected both the resources available and the kinds of knowledge that have been in demand. Yet government has played a secondary role. Indeed, one of the ironies of the situation is that many scholars who have strongly favored government central planning of economic activity have recognized very clearly the danger to scientific progress that would be imposed by central government planning of science, the danger of having priorities imposed from above rather than emerging spontaneously from the gropings and explorations of individual scientists.

A society's values, its culture, its social conventions—all these develop in the same way, through voluntary exchange, spontaneous cooperation, the evolution of a complex structure through trial and error, acceptance and rejection. No monarch ever decreed that the kind of music that is enjoyed by residents of Calcutta, for example, should differ radically from the kind enjoyed by residents of Vienna. These widely different musical cultures developed without anyone's "planning" them that way, through a kind of social evolution paralleling biological evolution—though, of course, individual sovereigns or even elected governments may have affected the direction of social evolution by sponsoring one or another musician or type of music, just as wealthy private individuals did.

The structures produced by voluntary exchange, whether they be language or scientific discoveries or musical styles or economic systems, develop a life of their own. They are capable of taking many different forms under different circumstances. Voluntary exchange can produce uniformity in some respects combined with diversity in others. It is a subtle process whose general principles of operation can fairly readily be grasped but whose detailed results can seldom be foreseen.

These examples may suggest not only the wide scope for voluntary exchange but also the broad meaning that must be attached to the concept of "self-interest." Narrow preoccupation with the economic market has led to a narrow interpretation of self-interest as myopic selfishness, as exclusive concern with immediate material rewards. Economics has been berated for allegedly drawing far-reaching conclusions from a wholly unrealistic "economic man" who is little more than a calculating machine, responding only to monetary stimuli. That is a great mistake. Self-interest is not myopic selfishness. It is whatever it is that interests the participants, whatever they value, whatever goals they pursue. The scientist seeking to advance the frontiers of his discipline, the missionary seeking to convert infidels to the true faith, the philanthropist seeking to bring comfort to the needy —all are pursuing their interests, as they see them, as they judge them by their own values.

THE ROLE OF GOVERNMENT

Where does government enter into the picture? To some extent government is a form of voluntary cooperation, a way in which people choose to achieve some of their objectives through governmental entities because they believe that is the most effective means of achieving them.

The clearest example is local government under conditions

where people are free to choose where to live. You may decide to live in one community rather than another partly on the basis of the kind of services its government offers. If it engages in activities you object to or are unwilling to pay for, and these more than balance the activities you favor and are willing to pay for, you can vote with your feet by moving elsewhere. There is competition, limited but real, so long as there are available alternatives.

But government is more than that. It is also the agency that is widely regarded as having a monopoly on the legitimate use of force or the threat of force as the means through which some of us can legitimately impose restraints through force upon others of us. The role of government in that more basic sense has changed drastically over time in most societies and has differed widely among societies at any given time. Much of the rest of this book deals with how its role has changed in the United States in recent decades, and what the effects of its activities have been.

In this initial sketch we want to consider a very different question. In a society whose participants desire to achieve the greatest possible freedom to choose as individuals, as families, as members of voluntary groups, as citizens of an organized government, what role should be assigned to government?

It is not easy to improve on the answer that Adam Smith gave to this question two hundred years ago:

> All systems either of preference or of restraint, therefore, being thus completely taken away, the obvious and simple system of natural liberty establishes itself of its own accord. Every man, as long as he does not violate the laws of justice, is left perfectly free to pursue his own interest his own way, and to bring both his industry and capital into competition with those of any other man, or order of men. The sovereign is completely discharged from a duty, in the attempting to perform which he must always be exposed to innumerable delusions, and for the proper performance of which no human wisdom or knowledge could ever be sufficient; the duty of superintending the industry of private people, and of directing it towards the employments most suitable to the interest of the society. According to the system of natural liberty, the sovereign has only three duties to attend to; three duties of great importance, indeed, but plain and intelligible to common understandings: first, the duty of protecting the society from the violence and invasion of other independent societies;

secondly, the duty of protecting, as far as possible, every member of the society from the injustice or oppression of every other member of it, or the duty of establishing an exact administration of justice; and, thirdly, the duty of erecting and maintaining certain public works and certain public institutions, which it can never be for the interest of any individual, or small number of individuals, to erect and maintain; because the profit could never repay the expence to any individual or small number of individuals, though it may frequently do much more than repay it to a great society.[3]

The first two duties are clear and straightforward: the protection of individuals in the society from coercion whether it comes from outside or from their fellow citizens. Unless there is such protection, we are not really free to choose. The armed robber's "Your money or your life" offers me a choice, but no one would describe it as a free choice or the subsequent exchange as voluntary.

Of course, as we shall see repeatedly throughout this book, it is one thing to state the purpose that an institution, particularly a governmental institution, "ought" to serve; it is quite another to describe the purposes the institution actually serves. The intentions of the persons responsible for setting up the institution and of the persons who operate it often differ sharply. Equally important, the results achieved often differ widely from those intended.

Military and police forces are required to prevent coercion from without and within. They do not always succeed and the power they possess is sometimes used for very different purposes. A major problem in achieving and preserving a free society is precisely how to assure that coercive powers granted to government in order to preserve freedom are limited to that function and are kept from becoming a threat to freedom. The founders of our country wrestled with that problem in drawing up the Constitution. We have tended to neglect it.

Adam Smith's second duty goes beyond the narrow police function of protecting people from physical coercion; it includes "an exact administration of justice." No voluntary exchange that is at all complicated or extends over any considerable period of time can be free from ambiguity. There is not enough fine print in the world to specify in advance every contingency that might arise and to describe precisely the obligations of the various parties to the exchange in each case. There must be some way to mediate disputes. Such mediation itself can be voluntary and need not involve government. In the United States today, most disagreements that arise in connection with commercial contracts are settled by resort to private arbitrators chosen by a procedure specified in advance. In response to this demand an extensive private judicial system has grown up. But the court of last resort is provided by the governmental judicial system.

This role of government also includes facilitating voluntary exchanges by adopting general rules—the rules of the economic and social game that the citizens of a free society play. The most obvious example is the meaning to be attached to private property. I own a house. Are you "trespassing" on my private property if you fly your private airplane ten feet over my roof? One thousand feet? Thirty thousand feet? There is nothing "natural" about where my property rights end and yours begin. The major way that society has come to agree on the rules of property is through the growth of common law, though more recently legislation has played an increasing role.

Adam Smith's third duty raises the most troublesome issues. He himself ˆegarded it as having a narrow application. It has since been used to justify an extremely wide range of government activities. In our view it describes a valid duty of a government directed to preserving and strengthening a free society; but it ¢an also be interpreted to justify unlimited extensions of government power.

The valid element arises because of the cost of producing some goods or services through strictly voluntary exchanges. To take one simple example suggested directly by Smith's description of the third duty: city streets and general-access highways could be provided by private voluntary exchange, the costs being paid for by charging tolls. But the costs of collecting the tolls would often be very large compared to the cost of building and maintaining the streets or highways. This is a "public work" that it might not "be for the interest of any individual . . . to erect and maintain . . . though it" might be worthwhile for "a great society."

A more subtle example involves effects on "third parties,"

people who are not parties to the particular exchange—the classic "smoke nuisance" case. Your furnace pours forth sooty smoke that dirties a third party's shirt collar. You have unintentionally imposed costs on a third party. He would be willing to let you dirty his collar for a price—but it is simply not feasible for you to identify all of the people whom you affect or for them to discover who has dirtied their collars and to require you to indemnify them individually or reach individual agreements with them.

The effect of your actions on third parties may be to confer benefits rather than impose costs. You landscape your house beautifully, and all passersby enjoy the sight. They would be willing to pay something for the privilege but it is not feasible to charge them for looking at your lovely flowers.

To lapse into technical jargon, there is a "market failure" because of "external" or "neighborhood" effects for which it is not feasible (i.e., would cost too much) to compensate or charge the people affected; third parties have had involuntary exchanges imposed on them.

Almost everything we do has some third-party effects, however small and however remote. In consequence, Adam Smith's third duty may at first blush appear to justify almost any proposed government measure. But there is a fallacy. Government measures also have third-party effects. "Government failure" no less than "market failure" arises from "external" or "neighborhood" effects. And if such effects are important for a market transaction, they are likely also to be important for government measures intended to correct the "market failure." The primary source of significant third-party effects of private actions is the difficulty of identifying the external costs or benefits. When it is easy to identify who is hurt or who is benefited, and by how much, it is fairly straightforward to replace involuntary by voluntary exchange, or at least to require individual compensation. If your car hits someone else's because of your negligence, you can be made to pay him for damages even though the exchange was involuntary. If it were easy to know whose collars were going to be dirtied, it would be possible for you to compensate the people affected, or alternatively, for them to pay you to pour out less smoke.

If it is difficult for private parties to identify who imposes costs or benefits on whom, it is difficult for government to do so. As a result a government attempt to rectify the situation may very well end up making matters worse rather than better—imposing costs on innocent third parties or conferring benefits on lucky bystanders. To finance its activities it must collect taxes, which themselves affect what the taxpayers do—still another third-party effect. In addition, every accretion of government power for whatever purpose increases the danger that government, instead of serving the great majority of its citizens, will become a means whereby some of its citizens can take advantage of others. Every government measure bears, as it were, a smokestack on its back.

Voluntary arrangements can allow for third-party effects to a much greater extent than may at first appear. To take a trivial example, tipping at restaurants is a social custom that leads you to assure better service for people you may not know or ever meet and, in return, be assured better service by the actions of still another group of anonymous third parties. Nonetheless, third-party effects of private actions do occur that are sufficiently important to justify government action. The lesson to be drawn from the misuse of Smith's third duty is not that government intervention is never justified, but rather that the burden of proof should be on its proponents. We should develop the practice of examining both the benefits and the costs of proposed government interventions and require a very clear balance of benefits over costs before adopting them. This course of action is recommended not only by the difficulty of assessing the hidden costs of government intervention but also by another consideration. Experience shows that once government undertakes an activity, it is seldom terminated. The activity may not live up to expectation but that is more likely to lead to its expansion, to its being granted a larger budget, than to its curtailment or abolition.

A fourth duty of government that Adam Smith did not explicitly mention is the duty to protect members of the community who cannot be regarded as "responsible" individuals. Like Adam Smith's third duty, this one, too, is susceptible of great abuse. Yet it cannot be avoided.

Freedom is a tenable objective only for responsible individuals. We do not believe in freedom for madmen or children. We must somehow draw a line between responsible individuals and others, yet doing so introduces a fundamental ambiguity into our ultimate objective of freedom. We cannot categorically reject paternalism for those whom we consider as not responsible.

For children we assign responsibility in the first instance to parents. The family, rather than the individual, has always been and remains today the basic building block of our society, though its hold has clearly been weakening—one of the most unfortunate consequences of the growth of government paternalism. Yet the assignment of responsibility for children to their parents is largely a matter of expediency rather than principle. We believe, and with good reason, that parents have more interest in their children than anyone else and can be relied on to protect them and to assure their development into responsible adults. However, we do not believe in the right of the parents to do whatever they will with their children—to beat them, murder them, or sell them into slavery. Children are responsible individuals in embryo. They have ultimate rights of their own and are not simply the playthings of their parents.

Adam Smith's three duties, or our four duties of government, are indeed "of great importance," but they are far less "plain and intelligible to common understandings" than he supposed. Though we cannot decide the desirability or undesirability of any actual or proposed government intervention by mechanical reference to one or another of them, they provide a set of principles that we can use in casting up a balance sheet of pros and cons. Even on the loosest interpretation, they rule out much existing government intervention—all those "systems either of preference or of restraint" that Adam Smith fought against, that were subsequently destroyed, but have since reappeared in the form of today's tariffs, governmentally fixed prices and wages, restrictions on entry into various occupations, and numerous other departures from his "simple system of natural liberty." (Many of these are discussed in later chapters.)

LIMITED GOVERNMENT IN PRACTICE

In today's world big government seems pervasive. We may well ask whether there exist any contemporaneous examples of societies that rely primarily on voluntary exchange through the market to organize their economic activity and in which government is limited to our four duties.

Perhaps the best example is Hong Kong—a speck of land next to mainland China containing less than 400 square miles with a population of roughly 4.5 million people. The density of population is almost unbelievable—14 times as many people per square mile as in Japan, 185 times as many as in the United States. Yet they enjoy one of the highest standards of living in all of Asia— second only to Japan and perhaps Singapore.

Hong Kong has no tariffs or other restraints on international trade (except for a few "voluntary" restraints imposed by. the United States and some other major countries). It has no government direction of economic activity, no minimum wage laws, no fixing of prices. The residents are free to buy from whom they want, to sell to whom they want, to invest however they want, to hire whom they want, to work for whom they want.

Government plays an important role that is limited primarily to our four duties interpreted rather narrowly. It enforces law and order, provides a means for formulating the rules of conduct, adjudicates disputes, facilitates transportation and communication, and supervises the issuance of currency. It has provided public housing for arriving refugees from China. Though government spending has grown as the economy has grown, it remains among the lowest in the world as a fraction of the income of the people. As a result, low taxes preserve incentives. Businessmen can reap the benefits of their success but must also bear the costs of their mistakes.

It is somewhat ironic that Hong Kong, a Crown colony of Great Britain, should be the modern exemplar of free markets and limited government. The British officials who govern it have enabled Hong Kong to flourish by following policies radically at variance with the welfare state policies that have been adopted by the mother country.

Though Hong Kong is an excellent current example, it is by no means the most important example of limited government and free market societies in practice. For this we must go back in time to the nineteenth century. One example, Japan in the first thirty years after the Meiji Restoration in 1867, we leave for Chapter 2.

Two other examples are Great Britain and the United States. Adam Smith's *Wealth of Nations* was one of the early blows in the battle to end government restrictions on industry and trade. The final victory in that battle came seventy years later, in 1846, with the repeal of the so-called Corn Laws—laws that imposed tariffs and other restrictions on the importation of wheat and other grains, referred to collectively as "corn." That ushered in three-quarters of a century of complete free trade lasting until the outbreak of World War I and completed a transition that had begun decades earlier to a highly limited government, one that left every resident of Britain, in Adam Smith's words quoted earlier, "perfectly free to pursue his own interest his own way, and to bring both his industry and capital into competition with those of any other man, or order of men."

Economic growth was rapid. The standard of life of the ordinary citizen improved dramatically—making all the more visible the remaining areas of poverty and misery portrayed so movingly by Dickens and other contemporary novelists. Population increased along with the standard of life. Britain grew in power and influence around the world. All this while government spending fell as a fraction of national income—from close to one-quarter of the national income early in the nineteenth century to about one-tenth of national income at the time of Queen Victoria's Jubilee in 1897, when Britain was at the very apex of its power and glory.

The United States is another striking example. There were tariffs, justified by Alexander Hamilton in his famous *Report on Manufactures* in which he attempted—with a decided lack of success—to refute Adam Smith's arguments in favor of free trade. But they were modest, by modern standards, and few other government restrictions impeded free trade at home or abroad. Until after World War I immigration was almost completely free (there were restrictions on immigration from the Orient). As the Statue of Liberty inscription has it:

> *Give me your tired, your poor,*
> *Your huddled masses yearning to breathe free,*
> *The wretched refuse of your teeming shore.*
> *Send these, the homeless, tempest-tossed to me:*
> *I lift my lamp beside the golden door.*

They came by the millions, and by the millions they were absorbed. They prospered because they were left to their own devices.

A myth has grown up about the United States that paints the nineteenth century as the era of the robber baron, of rugged, unrestrained individualism. Heartless monopoly capitalists allegedly exploited the poor, encouraged immigration, and then fleeced the immigrants unmercifully. Wall Street is pictured as conning Main Street, as bleeding the sturdy farmers in the Middle West, who survived despite the widespread distress and misery inflicted on them.

The reality was very different. Immigrants kept coming. The early ones might have been fooled, but it is inconceivable that millions kept coming to the United States decade after decade to be exploited. They came because the hopes of those who had preceded them were largely realized. The streets of New York were not paved with gold, but hard work, thrift, and enterprise brought rewards that were not even imaginable in the Old World. The newcomers spread from east to west. As they spread, cities sprang up, ever more land was brought into cultivation. The country grew more prosperous and more productive, and the immigrants shared in the prosperity.

If farmers were exploited, why did their number increase? The prices of farm products did decline. But that was a sign of success, not of failure, reflecting the development of machinery, the bringing under cultivation of more land, and improvements in communication, all of which led to a rapid growth in farm output. The final proof is that the price of farmland rose steadily—hardly a sign that farming was a depressed industry!

The charge of heartlessness, epitomized in the remark that William H. Vanderbilt, a railroad tycoon, is said to have made to an inquiring reporter, "The public be damned," is belied by the flowering of charitable activity in the United States in the nineteenth century. Privately financed schools and colleges multiplied; foreign missionary activity exploded; nonprofit private hospitals, orphanages, and numerous other institutions sprang up like weeds. Almost every charitable or public service organization, from the Society for the Prevention of Cruelty to Animals to the YMCA and YWCA, from the Indian Rights Association to the Salvation Army, dates from that period. Voluntary cooperation is no less effective in organizing charitable activity than in organizing production for profit.

The charitable activity was matched by a burst of cultural activity—art museums, opera houses, symphonies, museums, public libraries arose in big cities and frontier towns alike.

The size of government spending is one measure of government's role. Major wars aside, government spending from 1800 to 1929 did not exceed about 12 percent of the national income. Two-thirds of that was spent by state and local governments, mostly for schools and roads. As late as 1928, federal government spending amounted to about 3 percent of the national income.

The success of the United States is often attributed to its generous natural resources and wide open spaces. They certainly played a part—but then, if they were crucial, what explains the success of nineteenth-century Great Britain and Japan or twentieth-century Hong Kong?

It is often maintained that while a let-alone, limited government policy was feasible in sparsely settled nineteenth-century America, government must play a far larger, indeed dominant, role in a modern urbanized and industrial society. One hour in Hong Kong will dispose of that view.

Our society is what we make it. We can shape our institutions. Physical and human characteristics limit the alternatives available to us. But none prevents us, if we will, from building a society that relies primarily on voluntary cooperation to organize both economic and other activity, a society that preserves and expands human freedom, that keeps government in its place, keeping it our servant and not letting it become our master.

CHAPTER 2

The Tyranny
of Controls

In discussing tariffs and other restrictions on international trade in his *Wealth of Nations*, Adam Smith wrote:

> What is prudence in the conduct of every private family, can scarce be folly in that of a great kingdom. If a foreign country can supply us with a commodity cheaper than we ourselves can make it, better buy it of them with some part of the produce of our own industry, employed in a way in which we have some advantage. . . . In every country, it always is and must be the interest of the great body of the people to buy whatever they want of those who sell it cheapest. The proposition is so very manifest, that it seems ridiculous to take any pains to prove it; nor could it ever have been called in question, had not the interested sophistry of merchants and manufacturers confounded the common sense of mankind. Their interest is, in this respect, directly opposite to that of the great body of the people.[1]

These words are as true today as they were then. In domestic as well as foreign trade, it is in the interest of "the great body of the people" to buy from the cheapest source and sell to the dearest. Yet "interested sophistry" has led to a bewildering proliferation of restrictions on what we may buy and sell, from whom we may buy and to whom we may sell and on what terms, whom we may employ and whom we may work for, where we may live, and what we may eat and drink.

Adam Smith pointed to "the interested sophistry of merchants and manufacturers." They may have been the chief culprits in his day. Today they have much company. Indeed, there is hardly one of us who is not engaged in "interested sophistry" in one area or another. In Pogo's immortal words, "We have met the enemy and they is us." We rail against "special interests" except when the "special interest" happens to be our own. Each of us knows that what is good for him is good for the country—so *our* "special interest" is different. The end result is a maze of restraints and restrictions that makes almost all of us worse off than we would be if they were all eliminated. We lose far more from measures that serve other "special interests" than we gain from measures that serve our "special interest."

The clearest example is in international trade. The gains to some producers from tariffs and other restrictions are more than offset by the loss to other producers and especially to consumers in general. Free trade would not only promote our material welfare, it would also foster peace and harmony among nations and spur domestic competition.

Controls on foreign trade extend to domestic trade. They become intertwined with every aspect of economic activity. Such controls have often been defended, particularly for underdeveloped countries, as essential to provide development and progress. A comparison of the experience of Japan after the Meiji Restoration in 1867 and of India after independence in 1947 tests this view. It suggests, as do other examples, that free trade at home and abroad is the best way that a poor country can promote the well-being of its citizens.

The economic controls that have proliferated in the United States in recent decades have not only restricted our freedom to use our economic resources, they have also affected our freedom of speech, of press, and of religion.

INTERNATIONAL TRADE

It is often said that bad economic policy reflects disagreement among the experts; that if all economists gave the same advice, economic policy would be good. Economists often do disagree, but that has not been true with respect to international trade. Ever since Adam Smith there has been virtual unanimity among economists, whatever their ideological position on other issues, that international free trade is in the best interest of the trading countries and of the world. Yet tariffs have been the rule. The only major exceptions are nearly a century of free trade in Great Britain after the repeal of the Corn Laws in 1846, thirty years of free trade in Japan after the Meiji Restoration, and free trade in Hong Kong today. The United States had tariffs throughout the nineteenth century and they were raised still higher in the twentieth century, especially by the Smoot-Hawley tariff bill of 1930, which some scholars regard as partly responsible for the severity of the subsequent depression. Tariffs have since been reduced by repeated international agreements, but they remain high, probably higher than in the nineteenth century, though the vast changes in the kinds of items entering international trade make a precise comparison impossible.

Today, as always, there is much support for tariffs—euphemistically labeled "protection," a good label for a bad cause. Producers of steel and steelworkers' unions press for restrictions on steel imports from Japan. Producers of TV sets and their workers lobby for "voluntary agreements" to limit imports of TV sets or components from Japan, Taiwan, or Hong Kong. Producers of textiles, shoes, cattle, sugar—they and myriad others complain about "unfair" competition from abroad and demand that government do something to "protect" them. Of course, no group makes its claim on the basis of naked self-interest. Every group speaks of the "general interest," of the need to preserve jobs or to promote national security. The need to strengthen the dollar vis-à-vis the mark or the yen has more recently joined the traditional rationalizations for restrictions on imports.

The Economic Case for Free Trade

One voice that is hardly ever raised is the consumer's. So-called consumer special interest groups have proliferated in recent years. But you will search the news media, or the records of congressional hearings in vain, to find any record of their launching a concentrated attack on tariffs or other restrictions on imports, even though consumers are major victims of such measures. The self-styled consumer advocates have other concerns—as we shall see in Chapter 7.

The individual consumer's voice is drowned out in the cacophony of the "interested sophistry of merchants and manufacturers" and their employees. The result is a serious distortion of the issue. For example, the supporters of tariffs treat it as self-evident that the creation of jobs is a desirable end, in and of itself, regardless of what the persons employed do. That is clearly wrong. If all we want are jobs, we can create any number—for example, have people dig holes and then fill them up again, or perform other useless tasks. Work is sometimes its own reward. Mostly, however, it is the price we pay to get the things we want. Our real objective is not just jobs but productive jobs—jobs that will mean more goods and services to consume.

Another fallacy seldom contradicted is that exports are good, imports bad. The truth is very different. We cannot eat, wear, or enjoy the goods we send abroad. We eat bananas from Central America, wear Italian shoes, drive German automobiles, and enjoy programs we see on our Japanese TV sets. Our gain from foreign trade is what we import. Exports are the price we pay to get imports. As Adam Smith saw so clearly, the citizens of a nation benefit from getting as large a volume of imports as possible in return for its exports, or equivalently, from exporting as little as possible to pay for its imports.

The misleading terminology we use reflects these erroneous ideas. "Protection" really means exploiting the consumer. A "favorable balance of trade" really means exporting more than we import, sending abroad goods of greater total value than the goods we get from abroad. In your private household, you would surely prefer to pay less for more rather than the other way around, yet that would be termed an "unfavorable balance of payments" in foreign trade.

The argument in favor of tariffs that has the greatest emotional appeal to the public at large is the alleged need to protect the high standard of living of American workers from the "unfair" competition of workers in Japan or Korea or Hong Kong who are willing to work for a much lower wage. What is wrong with this argument? Don't we want to protect the high standard of living of our people?

The fallacy in this argument is the loose use of the terms "high" wage and "low" wage. What do high and low wages mean? American workers are paid in dollars; Japanese workers are paid in yen. How do we compare wages in dollars with wages in yen? How many yen equal a dollar? What determines that exchange rate?

Consider an extreme case. Suppose that, to begin with, 360 yen equal a dollar. At this exchange rate, the actual rate of exchange for many years, suppose that the Japanese can produce and sell everything for fewer dollars than we can in the United States—TV sets, automobiles, steel, and even soybeans, wheat, milk, and ice cream. If we had free international trade, we would try to buy all our goods from Japan. This would seem to be the extreme horror story of the kind depicted by defenders of tariffs—we would be flooded with Japanese goods and could sell them nothing.

Before throwing up your hands in horror, carry the analysis one step further. How would we pay the Japanese? We would offer them dollar bills. What would they do with the dollar bills? We have assumed that at 360 yen to the dollar everything is cheaper in Japan, so there is nothing in the U.S. market that they would want to buy. If the Japanese exporters were willing to burn or bury the dollar bills, that would be wonderful for us. We would get all kinds of goods for green pieces of paper that we can produce in great abundance and very cheaply. We would have the most marvelous export industry conceivable.

Of course, the Japanese would not in fact sell us useful goods in order to get useless pieces of paper to bury or burn. Like us, they want to get something real in return for their work. If all goods were cheaper in Japan than in the United States at 360 yen to the dollar, the exporters would try to get rid of their dollars, would try to sell them for 360 yen to the dollar in order to buy the cheaper Japanese goods. But who would be willing to buy the dollars? What is true for the Japanese exporter is true for everyone in Japan. No one will be willing to give 360 yen in exchange for one dollar if 360 yen will buy more of everything in Japan than one dollar will buy in the United States. The exporters, on discovering that no one will buy their dollars at 360 yen, will offer to take fewer yen for a dollar. The price of the dollar in terms of yen will go down—to 300 yen for a dollar, or 250 yen, or 200 yen. Put the other way around, it will take more and more dollars to buy a given number of Japanese yen. Japanese goods are priced in yen, so their price in dollars will go up. Conversely, U.S. goods are priced in dollars, so the more dollars the Japanese get for a given number of yen, the cheaper U.S. goods become to the Japanese in terms of yen.

The price of the dollar in terms of yen would fall until, on the average, the dollar value of goods that the Japanese buy from the United States roughly equaled the dollar value of goods that the United States buys from Japan. At that price everybody who wanted to buy yen for dollars would find someone who was willing to sell him yen for dollars.

The actual situation is, of course, more complicated than this hypothetical example. Many nations, and not merely the United States and Japan, are engaged in trade, and the trade often takes roundabout directions. The Japanese may spend some of the dollars they earn in Brazil, the Brazilians in turn may spend those dollars in Germany, and the Germans in the United States, and so on in endless complexity. However, the principle is the same. People, in whatever country, want dollars primarily to buy useful items, not to hoard.

Another complication is that dollars and yen are used not only to buy goods and services from other countries but also to invest and make gifts. Throughout the nineteenth century the United States had a balance of payments deficit almost every year—an "unfavorable" balance of trade that was good for everyone. Foreigners wanted to invest capital in the United States. The British, for example, were producing goods and sending them to us in return for pieces of paper—not dollar bills, but bonds promising to pay back a sum of money at a later time plus interest. The British were willing to send us their goods because they regarded those bonds as a good investment. On the average, they were right. They received a higher return on their savings than was available in any other way. We, in turn, benefited by foreign investment that enabled us to develop more rapidly than we could have developed if we had been forced to rely solely on our own savings.

In the twentieth century the situation was reversed. U.S. citizens found that they could get a higher return on their capital by investing abroad than they could at home. As a result the United States sent goods abroad in return for evidence of debt—bonds and the like. After World War II, the U.S. government made gifts abroad in the form of the Marshall Plan and other foreign aid programs. We sent goods and services abroad as an expression of our belief that we were thereby contributing to a more peaceful world. These government gifts supplemented private gifts—from charitable groups, churches supporting missionaries, individuals contributing to the support of relatives abroad, and so on.

None of these complications alters the conclusion suggested by the hypothetical extreme case. In the real world, as well as in that hypothetical world, there can be no balance of payments problem so long as the price of the dollar in terms of the yen or the mark or the franc is determined in a free market by voluntary transactions. It is simply not true that high-wage American workers are, as a group, threatened by "unfair" competition from low-wage foreign workers. Of course, particular workers may be harmed if a new or improved product is developed abroad, or if foreign producers become able to produce such products more cheaply. But that is no different from the effect on a particular group of workers of other American firms' developing new or improved products or discovering how to produce at lower costs. That is simply market competition in practice, the major source of the high standard of life of the American worker. If we want to benefit from a vital, dynamic, innovative economic system, we must accept the need for mobility and adjustment. It may be desirable to ease these adjustments, and we have adopted many arrangements, such as unemployment insurance, to do so, but we should try to achieve that objective without destroying the flexibility of the system— that would be to kill the goose that has been laying the golden eggs. In any event, whatever we do should be evenhanded with respect to foreign and domestic trade.

What determines the items it pays us to import and to export? An American worker is currently more productive than a Japanese worker. It is hard to determine just how much more productive— estimates differ. But suppose he is one and a half times as productive. Then, on average, the American's wages would buy about one and a half times as much as a Japanese worker's wages. It is wasteful to use American workers to do anything at which they are less than one and a half times as efficient as their Japanese counterparts. In the economic jargon coined more than 150 years ago, that is the *principle of comparative advantage*. Even if we were more efficient than the Japanese at producing everything, it would not pay us to produce everything. We should concentrate on doing those things we do best, those things where our superiority is the greatest. As a homely illustration, should a lawyer who can type twice as fast as his secretary fire the secretary and do his own typing? If the lawyer is twice as good a typist but five times as good a lawyer as his secretary, both he and the secretary are better off if he practices law and the secretary types letters.

Another source of "unfair competition" is said to be subsidies by foreign governments to their producers that enable them to sell in the United States below cost. Suppose a foreign government gives such subsidies, as no doubt some do. Who is hurt and who benefits? To pay for the subsidies the foreign government must tax its citizens. They are the ones who pay for the subsidies. U.S. consumers benefit. They get cheap TV sets or automobiles or whatever it is that is subsidized. Should we complain about such a program of reverse foreign aid? Was it noble of the United States to send goods and services as gifts to other countries in the form of Marshall Plan aid or, later, foreign aid, but ignoble for foreign countries to send us gifts in the indirect form of goods and services sold to us below cost? The citizens of the foreign government might well complain. They must suffer a lower standard of living for the benefit of American consumers and of some of their fellow citizens who own or work in the industries that are subsidized. No doubt, if such subsidies are introduced suddenly or erratically, that will adversely affect owners and workers in U.S. industries producing the same products. However, that is one of the ordinary risks of doing business. Enterprises never complain about unusual or accidental events that confer windfall gains. The free enterprise system is a *profit* and *loss* system. As already noted, any measures to ease the adjustment to sudden changes should be applied evenhandedly to domestic and foreign trade.

In any event, disturbances are likely to be temporary. Suppose that, for whatever reason, Japan decided to subsidize steel very heavily. If no additional tariffs or quotas were imposed, imports of steel into the United States would go up sharply. That would drive down the price of steel in the United States and force steel producers to cut their output, causing unemployment in the steel industry. On the other hand, products made of steel could be purchased more cheaply. Buyers of such products would have extra money to spend on other things. The demand for other items would go up, as would employment in enterprises producing those items. Of course, it would take time to absorb the now unemployed steelworkers. However, to balance that effect, workers in other industries who had been unemployed would find jobs available. There need be no net loss of employment, and there would be a gain in output because workers no longer needed to produce steel would be available to produce something else.

The same fallacy of looking at only one side of the issue is present when tariffs are urged in order to add to employment. If tariffs are imposed on, say, textiles, that will add to output and employment in the domestic textile industry. However, foreign producers who no longer can sell their textiles in the United States earn fewer dollars. They will have less to spend in the United States. Exports will go down to balance decreased imports. Employment will go up in the textile industry, down in the export industries. And the shift of employment to less productive uses will reduce total output.

The national security argument that a thriving domestic steel industry, for example, is needed for defense has no better basis. National aefense needs take only a small fraction of total steel used in the United States. And it is inconceivable that complete free trade in steel would destroy the U.S. steel industry. The advantages of being close to sources of supply and fuel and to the market would guarantee a relatively large domestic steel industry. Indeed, the need to meet foreign competition, rather than being sheltered behind governmental barriers, might very well produce a stronger and more efficient steel industry than we have today.

Suppose the improbable did happen. Suppose it did prove cheaper to buy *all* our steel abroad. There are alternative ways to provide for national security. We could stockpile steel. That is easy, since steel takes relatively little space and is not perishable. We could maintain some steel plants in mothballs, the way we maintain ships, to go into production in case of need. No doubt there are still other alternatives. Before a steel company decides to build a new plant, it investigates alternative ways of doing so, alternative locations, in order to choose the most efficient and economical. Yet in all its pleas for subsidies on national security grounds, the steel industry has never presented cost estimates for alternative ways of providing national security. Until they do, we can be sure the national security argument is a rationalization of industry self-interest, not a valid reason for the subsidies.

No doubt the executives of the steel industry and of the steel labor unions are sincere when they adduce national security arguments. Sincerity is a much overrated virtue. We are all capable of persuading ourselves that what is good for us is good for the country. We should not complain about steel producers making such arguments, but about letting ourselves be taken in by them.

What about the argument that we must defend the dollar, that we must keep it from falling in value in terms of other currencies—the Japanese yen, the German mark, or the Swiss franc? That is a wholly artificial problem. If foreign exchange rates are determined in a free market, they will settle at whatever level will clear the market. The resulting price of the dollar in terms of the yen, say, may temporarily fall below the level justified by the cost in dollars and yen respectively of American and Japanese goods. If so, it will give persons who recognize that situation an incentive to buy dollars and hold them for a while in order to make a profit when the price goes up. By lowering the price in yen of American exports to Japanese, it will stimulate American exports; by raising the price in dollars of Japanese goods, it will discourage imports from Japan. These developments will increase the demand for dollars and so correct the initially low price. The price of the dollar, if determined freely, serves the same function as all other prices. It transmits information and provides an incentive to act on that information because it affects the incomes that participants in the market receive.

Why then all the furor about the "weakness" of the dollar? Why the repeated foreign exchange crises? The proximate reason is because foreign exchange rates have not been determined in a free market. Government central banks have intervened on a grand scale in order to influence the price of their currencies. In the process they have lost vast sums of their citizens' money (for the United States close to $2 billion from 1973 to early 1979). Even more important, they have prevented this important set of prices from performing its proper function. They have not been able to prevent the basic underlying economic forces from ultimately having their effect on exchange rates, but have been able to maintain artificial exchange rates for substantial intervals. The effect has been to prevent gradual adjustment to the underlying forces. Small disturbances have accumulated into large ones, and ultimately there has been a major foreign exchange "crisis."

Why have governments intervened in foreign exchange markets? Because foreign exchange rates reflect internal policies. The U.S. dollar has been weak compared to the Japanese yen, the German mark, and the Swiss franc primarily because inflation has been much higher in the United States than in the other countries. Inflation meant that the dollar was able to buy less and less at home. Should we be surprised that it has also been able to buy less abroad? Or that Japanese or Germans or Swiss should not be willing to exchange as many of their own currency units for a dollar? But governments, like the rest of us, go to great lengths to try to conceal or offset the undesirable consequences of their own policies. A government that inflates is therefore led to try to manipulate the foreign exchange rate. When it fails, it blames internal inflation on the decline in the exchange rate, instead of acknowledging that cause and effect run the other way.

In all the voluminous literature of the past several centuries on free trade and protectionism, only three arguments have ever been advanced in favor of tariffs that even in principle may have some validity.

First is the national security argument already mentioned. Although that argument is more often a rationalization for particular tariffs than a valid reason for them, it cannot be denied that on occasion it might justify the maintenance of otherwise uneconomical productive facilities. To go beyond this statement of possibility and establish in a specific case that a tariff or other trade restriction is justified in order to promote national security, it would be necessary to compare the cost of achieving the specific security objective in alternative ways and establish at least a *prima facie* case that a tariff is the least costly way. Such cost comparisons are seldom made in practice.

The second is the "infant industry" argument advanced, for

example, by Alexander Hamilton in his *Report on Manufactures*. There is, it is said, a potential industry which, if once established and assisted during its growing pains, could compete on equal terms in the world market. A temporary tariff is said to be justified in order to shelter the potential industry in its infancy and enable it to grow to maturity, when it can stand on its own feet. Even if the industry could compete successfully once established, that does not of itself justify an initial tariff. It is worthwhile for consumers to subsidize the industry initially—which is what they in effect do by levying a tariff—only if they will subsequently get back at least that subsidy in some other way, through prices later lower than the world price, or through some other advantages of having the industry. But in that case, is a subsidy needed? Will it then not pay the original entrants into the industry to suffer initial losses in the expectation of being able to recoup them later? After all, most firms experience losses in their early years, when they are getting established. That is true if they enter a new industry or if they enter an existing one. Perhaps there may be some special reason why the original entrants cannot recoup their initial losses even though it be worthwhile for the community at large to make the initial investment. But surely the presumption is the other way.

The infant industry argument is a smoke screen. The so-called infants never grow up. Once imposed, tariffs are seldom eliminated. Moreover, the argument is seldom used on behalf of true unborn infants that might conceivably be born and survive if given temporary protection. They have no spokesmen. It is used to justify tariffs for rather aged infants that can mount political pressure.

The third argument for tariffs that cannot be dismissed out of hand is the "beggar-thy-neighbor" argument. A country that is a major producer of a product, or that can join with a small number of other producers that together control a major share of production, may be able to take advantage of its monopoly position by raising the price of the product (the OPEC cartel is the obvious current example). Instead of raising the price directly, the country can do so indirectly by imposing an export tax on the product—an export tariff. The benefit to itself will be less than the cost to others, but from the national point of view, there can be a gain. Similarly, a country that is the primary purchaser of a product—in economic jargon, has monopsony power—may be able to benefit by driving a hard bargain with the sellers and imposing an unduly low price on them. One way to do so is to impose a tariff on the import of the product. The net return to the seller is the price less the tariff, which is why this can be equivalent to buying at a lower price. In effect, the tariff is paid by the foreigners (we can think of no actual example). In practice this nationalistic approach is highly likely to promote retaliation by other countries. In addition, as for the infant industry argument, the actual political pressures tend to produce tariff structures that do not in fact take advantage of any monopoly or monopsony positions.

A fourth argument, one that was made by Alexander Hamilton and continues to be repeated down to the present, is that free trade would be fine if all other countries practiced free trade but that so long as they do not, the United States cannot afford to. This argument has no validity whatsoever, either in principle or in practice. Other countries that impose restrictions on international trade do hurt us. But they also hurt themselves. Aside from the three cases just considered, if we impose restrictions in turn, we simply add to the harm to ourselves and also harm them as well. Competition in masochism and sadism is hardly a prescription for sensible international economic policy! Far from leading to a reduction in restrictions by other countries, this kind of retaliatory action simply leads to further restrictions.

We are a great nation, the leader of the free world. It ill behooves us to require Hong Kong and Taiwan to impose export quotas on textiles to "protect" our textile industry at the expense of U.S. consumers and of Chinese workers in Hong Kong and Taiwan. We speak glowingly of the virtues of free trade, while we use our political and economic power to induce Japan to restrict exports of steel and TV sets. We should move unilaterally to free trade, not instantaneously, but over a period of, say, five years, at a pace announced in advance.

Few measures that we could take would do more to promote the cause of freedom at home and abroad than complete free trade. Instead of making grants to foreign governments in the name of economic aid—thereby promoting socialism—while at the same time imposing restrictions on the products they produce —thereby hindering free enterprise—we could assume a consistent and principled stance. We could say to the rest of the world: we believe in freedom and intend to practice it. We cannot force you to be free. But we can offer full cooperation on equal terms to all. Our market is open to you without tariffs or other restrictions. Sell here what you can and wish to. Buy whatever you can and wish to. In that way cooperation among individuals can be world-wide and free.

The Political Case for Free Trade

Interdependence is a pervasive characteristic of the modern world: in the economic sphere proper, between one set of prices and another, between one industry and another, between one country and another; in the broader society, between economic activity and cultural, social, and charitable activities; in the organization of society, between economic arrangements and political arrangements, between economic freedom and political freedom.

In the international sphere as well, economic arrangements are intertwined with political arrangements. International free trade fosters harmonious relations among nations that differ in culture and institutions just as free trade at home fosters harmonious relations among individuals who differ in beliefs, attitudes, and interests.

In a free trade world, as in a free economy in any one country, transactions take place among private entities—individuals, business enterprises, charitable organizations. The terms at which any transaction takes place are agreed on by all the parties to that transaction. The transaction will not take place unless all parties believe they will benefit from it. As a result, the interests of the various parties are harmonized. Cooperation, not conflict, is the rule.

When governments intervene, the situation is very different. Within a country, enterprises seek subsidies from their government, either directly or in the form of tariffs or other restrictions on trade. They will seek to evade economic pressures from competitors that threaten their profitability or their very existence by resorting to political pressure to impose costs on others. Intervention by one government in behalf of local enterprises leads enterprises in other countries to seek the aid of their own government to counteract the measures taken by the foreign government. Private disputes become the occasion for disputes between governments. Every trade negotiation becomes a political matter. High government officials jet around the world to trade conferences. Frictions develop. Many citizens of every country are disappointed at the outcome and end up feeling they got the short end of the stick. Conflict, not cooperation, is the rule.

The century from Waterloo to the First World War offers a striking example of the beneficial effects of free trade on the relations among nations. Britain was the leading nation of the world, and during the whole of that century it had nearly complete free trade. Other nations, particularly Western nations, including the United States, adopted a similar policy, if in somewhat diluted form. People were in the main free to buy and sell goods from and to anyone, wherever he lived, whether in the same or a different country, at whatever terms were mutually agreeable. Perhaps even more surprising to us today, people were free to travel all over Europe and much of the rest of the world without a passport and without repeated customs inspection. They were free to emigrate and in much of the world, particularly the United States, free to enter and become residents and citizens.

As a result, the century from Waterloo to the First World War was one of the most peaceful in human history among Western nations, marred only by some minor wars—the Crimean War and the Franco-Prussian Wars are the most memorable—and, of course, a major civil war within the United States, which itself was a result of the major respect—slavery—in which the United States departed from economic and political freedom.

In the modern world, tariffs and similar restrictions on trade have been one source of friction among nations. But a far more troublesome source has been the far-reaching intervention of the state into the economy in such collectivist states as Hitler's Germany, Mussolini's Italy, and Franco's Spain, and especially the communist countries, from Russia and its satellites to China. Tariffs and similar restrictions distort the signals transmitted by the price system, but at least they leave individuals free to respond to those distorted signals. The collectivist countries have introduced much farther-reaching command elements. Completely private transactions are impossible between citizens of a largely market economy and of a collectivist state. One side is necessarily represented by government officials. Political considerations are unavoidable, but friction would be minimized if the governments of market economies permitted their citizens the maximum possible leeway to make their own deals with collectivist governments. Trying to use trade as a political weapon or political measures as a means to increase trade with collectivist countries only makes the inevitable political frictions even worse.

Free International Trade and Internal Competition

The extent of competition at home is closely related to international trade arrangements. A public outcry against "trusts" and "monopolies" in the late nineteenth century led to the establishment of the Interstate Commerce Commission and the adoption of the Sherman Anti-Trust Law, later supplemented by many other legislative actions to promote competition. These measures have had very mixed effects. They have contributed in some ways to increased competition, but in others they have had perverse effects.

But no such measure, even if it lived up to every expectation of its sponsors, could do as much to assure effective competition as the elimination of all barriers to international trade. The existence of only three major automobile producers in the United States— and one of those on the verge of bankruptcy—does raise a threat of monopoly pricing. But let the automobile producers *of the world* compete with General Motors, Ford, and Chrysler for the custom of the American buyer, and the specter of monopoly pricing disappears.

So it is throughout. A monopoly can seldom be established within a country without overt and covert government assistance in the form of a tariff or some other device. It is close to impossible to do so on a world scale. The De Beers diamond monopoly is the only one we know of that appears to have succeeded. We know of no other that has been able to exist for long without the direct assistance of governments—the OPEC cartel and earlier rubber and coffee cartels being perhaps the most prominent examples. And most such government-sponsored cartels have not lasted long. They have broken down under the pressure of international competition—a fate that we believe awaits OPEC as well. In a world of free trade, international cartels would disappear even more quickly. Even in a world of trade restrictions, the United States, by free trade, unilateral if necessary, could come close to eliminating any danger of significant internal monopolies.

CENTRAL ECONOMIC PLANNING

Traveling in underdeveloped countries, we have over and over again been deeply impressed by the striking contrast between the ideas about facts held by the intellectuals of those countries and many intellectuals in the West, and the facts themselves.

Intellectuals everywhere take for granted that free enterprise capitalism and a free market are devices for exploiting the masses, while central economic planning is the wave of the future that will set their countries on the road to rapid economic progress. We shall not soon forget the tongue-lashing one of us received from a prominent, highly successful, and extremely literate Indian entrepreneur—physically the very model of the Marxist caricature of an obese capitalist—in reaction to remarks that he correctly interpreted as criticism of India's detailed central planning. He informed us in no uncertain terms that the government of a country as poor as India simply had to control imports, domestic production, and the allocation of investment—and by implication grant him the special privileges in all these areas that are the source of his own affluence—in order to assure that *social* priorities override the selfish demands of individuals. And he was simply echoing the views of the professors and other intellectuals in India and elsewhere.

The facts themselves are very different. Wherever we find any large element of individual freedom, some measure of progress in the material comforts at the disposal of ordinary citizens, and widespread hope of further progress in the future, there we also find that economic activity is organized mainly through the free market. Wherever the state undertakes to control in detail the economic activities of its citizens, wherever, that is, detailed central economic planning reigns, there ordinary citizens are in political fetters, have a low standard of living, and have little power to control their own destiny. The state may prosper and produce impressive monuments. Privileged classes may enjoy a full measure of material comforts. But the ordinary citizens are instruments to be used for the state's purposes, receiving no more than necessary to keep them docile and reasonably productive.

The most obvious example is the contrast between East and West Germany, originally part of one whole, torn asunder by the vicissitudes of warfare. People of the same blood, the same civilization, the same level of technical skill and knowledge inhabit the two parts. Which has prospered? Which had to erect a wall to pen in its citizens? Which must man it today with armed guards, assisted by fierce dogs, minefields, and similar devices of devilish ingenuity in order to frustrate brave and desperate citizens who are willing to risk their lives to leave their communist paradise for the capitalist hell on the other side of the wall?

On one side of that wall the brightly lit streets and stores are filled with cheerful, bustling people. Some are shopping for goods from all over the globe. Others are going to the numerous movie houses or other places of entertainment. They can buy freely newspapers and magazines expressing every variety of opinion. They speak with one another or with strangers on any subject and express a wide range of opinions without a single backward glance over the shoulder. A walk of a few hundred feet, after an hour spent in line, filling in forms and waiting for passports to be returned, will take you, as it took us, to the other side of that wall. There, the streets appear empty; the city, gray and pallid; the store windows, dull; the buildings, grimy. Wartime destruction has not yet been repaired after more than three decades. The only sign of cheerfulness or activity that we found during our brief visit to East Berlin was at the entertainment center. One hour in East Berlin is enough to understand why the authorities put up the wall.

It seemed a miracle when West Germany—a defeated and devastated country—became one of the strongest economies on the continent of Europe in less than a decade. It was the miracle of a free market. Ludwig Erhard, an economist, was the German Minister of Economics. On Sunday, the twentieth of June, 1948, he simultaneously introduced a new currency, today's Deutsche mark, and abolished almost all controls on wages and prices. He acted on a Sunday, he was fond of saying, because the offices of the French, American, and British occupation authorities were closed that day. Given their favorable attitudes toward controls, he was sure that if he had acted when the offices were open, the occupation authorities would have countermanded his orders. His measures worked like a charm. Within days the shops were full of goods. Within months the German economy was humming away.

Even two communist countries, Russia and Yugoslavia, offer a similar, though less extreme, contrast. Russia is closely controlled from the center. It has not been able to dispense wholly with private property and free markets, but it has tried to limit their scope as much as possible. Yugoslavia started down the same road. However, after Yugoslavia under Tito broke with Stalin's Russia, it changed its course drastically. It is still communist but deliberately promotes decentralization and the use of market forces. Most agricultural land is privately owned, its produce sold on relatively free markets. Small enterprises (those that have fewer than five employees) may be privately owned and operated. They are flourishing, particularly in handicrafts and tourism. Larger enterprises are workers' cooperatives—an inefficient form of organization but one that at least provides some opportunity for individual responsibility and initiative. The inhabitants of Yugoslavia are not free. They have a much lower standard of living than the inhabitants of neighboring Austria or other similar Western countries. Yet Yugoslavia strikes the observant traveler who comes to it from Russia, as we did, as a paradise by comparison.

In the Middle East, Israel, despite an announced socialist philosophy and policy and extensive government intervention into the economy, has a vigorous market sector, primarily as an in-

direct consequence of the importance of foreign trade. Its socialist policies have retarded its economic growth, yet its citizens
enjoy both more political freedom and a far higher standard of
living than the citizens of Egypt, which has suffered from a much
more extensive centralization of political power and which has
imposed much more rigid controls on economic activity.

In the Far East, Malaysia, Singapore, Korea, Taiwan, Hong
Kong, and Japan—all relying extensively on private markets—
are thriving. Their people are full of hope. An economic explosion is under way in these countries. As best such things can be
measured, the annual income per person in these countries in the
late 1970s ranged from about $700 in Malaysia to about $5,000
in Japan. By contrast, India, Indonesia, and Communist China,
all relying heavily on central planning, have experienced economic stagnation and political repression. The annual income
per person in those countries was less than $250.

The intellectual apologists for centralized economic planning
sang the praises of Mao's China until Mao's successors trumpeted
China's backwardness and bemoaned the lack of progress during
the past twenty-five years. Part of their design to modernize the
country is to let prices and markets play a larger role. These
tactics may produce sizable gains from the country's present low
economic level—as they did in Yugoslavia. However, the gains
will be severely limited so long as political control over economic
activity remains tight and private property is narrowly limited.
Moreover, letting the genie of private initiative out of the bottle
even to this limited extent will give rise to political problems that,
sooner or later, are likely to produce a reaction toward greater
authoritarianism. The opposite outcome, the collapse of communism and its replacement by a market system, seems far less
likely, though as incurable optimists, we do not rule it out completely. Similarly, once the aged Marshal Tito dies, Yugoslavia
will experience political instability that may produce a reaction
toward greater authoritarianism or, far less likely, a collapse of
existing collectivist arrangements.

An especially illuminating example, worth examining in greater
detail, is the contrast between the experiences of India and Japan
—India during the first thirty years after it achieved independence

in 1947, and Japan not today but during the first thirty years after the Meiji Restoration in 1867. Economists and social scientists in general can seldom conduct controlled experiments of the kind that are so important in testing hypotheses in the physical sciences. However, experience has here produced something very close to a controlled experiment that we can use to test the importance of the difference in methods of economic organization.

There is a lapse of eight decades in time. In all other respects the two countries were in very similar circumstances at the outset of the periods we compare. Both were countries with ancient civilizations and a sophisticated culture. Each had a highly structured population. Japan had a feudal structure with daimyos (feudal lords) and serfs. India had a rigid caste system with Brahmans at the top and the untouchables, designated by the British the "scheduled castes," at the bottom.

Both countries experienced a major political change that permitted a drastic alteration in political, economic, and social arrangements. In both countries a group of able, dedicated leaders took power. They were imbued with national pride and determined to convert economic stagnation into rapid growth, to transform their countries into great powers.

Almost all differences favored India rather than Japan. The prior rules of Japan had enforced almost complete isolation from the rest of the world. International trade and contact was limited to one visit from one Dutch ship a year. The few Westerners permitted to stay in the country were confined to a small enclave on an island in the harbor of Osaka. Three or more centuries of enforced isolation had left Japan ignorant of the outside world, far behind the West in science and technology, and with almost no one who could speak or read any foreign language other than Chinese.

India was much more fortunate. It had enjoyed substantial economic growth before World War I. That growth was converted into stagnation between the two world wars by the struggle for independence from Britain, but was not reversed. Improvements in transportation had ended the localized famines that had earlier been a recurrent curse. Many of its leaders had been educated in advanced countries of the West, particularly in Great

Britain. British rule left it with a highly skilled and trained civil service, modern factories, and an excellent railroad system. None of these existed in Japan in 1867. India was technologically backward compared to the West, but the differential was less than that between Japan in 1867 and the advanced countries of that day. India's physical resources, too, were far superior to Japan's. About the only physical advantage Japan had was the sea, which offered easy transportation and a plentiful supply of fish. For the rest, India is nearly nine times as large as Japan, and a much larger percentage of its area consists of relatively level and accessible land. Japan is mostly mountainous. It has only a narrow fringe of habitable and arable land along the seacoasts.

Finally, Japan was on its own. No foreign capital was invested in Japan; no foreign governments or foreign foundations in capitalist countries formed consortiums to make grants or offer low-interest loans to Japan. It had to depend on itself for capital to finance its economic development. It did have one lucky break. In the early years after the Meiji Restoration, the European silk crops experienced a disastrous failure that enabled Japan to earn more foreign exchange by silk exports than she otherwise could have. Aside from that, there were no important fortuitous or organized sources of capital.

India fared far better. Since it achieved independence in 1947, it has received an enormous volume of resources from the rest of the world, mostly as gifts. The flow continues today.

Despite the similar circumstances of Japan in 1867 and India in 1947, the outcome was vastly different. Japan dismantled its feudal structure and extended social and economic opportunity to all its citizens. The lot of the ordinary man improved rapidly, even though population exploded. Japan became a power to be reckoned with on the international political scene. It did not achieve full individual human and political freedom, but it made great progress in that direction.

India paid lip service to the elimination of caste barriers yet made little progress in practice. Differences in income and wealth between the few and the many grew wider, not narrower. Population exploded, as it did in Japan eight decades earlier, but economic output per capita did not. It remained nearly stationary.

Indeed, the standard of life of the poorest third of the population has probably declined. In the aftermath of British rule, India prided itself on being the largest democracy in the world, but it lapsed for a time into a dictatorship that restricted freedom of speech and press. It is in danger of doing so again.

What explains the difference in results? Many observers point to different social institutions and human characteristics. Religious taboos, the caste system, a fatalistic philosophy—all these are said to imprison the inhabitants of India in a straitjacket of tradition. The Indians are said to be unenterprising and slothful. By contrast, the Japanese are lauded as hardworking, energetic, eager to respond to influences from abroad, and incredibly ingenious at adapting what they learn from outside to their own needs.

This description of the Japanese may be accurate today. It was not in 1867. An early foreign resident in Japan wrote: "Wealthy we do not think it [Japan] will ever become. The advantages conferred by Nature, with exception of the climate, and the love of indolence and pleasure of the people themselves forbid it. The Japanese are a happy race, and being content with little are not likely to achieve much." Wrote another: "In this part of the world, principles, established and recognized in the West, appear to lose whatever virtue and vitality they originally possessed and to tend fatally toward weediness and corruption."

Similarly, the description of the Indians may be accurate today for some Indians in India, even perhaps for most, but it certainly is not accurate for Indians who have migrated elsewhere. In many African countries, in Malaya, Hong Kong, the Fiji Islands, Panama, and, most recently, Great Britain, Indians are successful entrepreneurs, sometimes constituting the mainstay of the entrepreneurial class. They have often been the dynamo initiating and promoting economic progress. Within India itself, enclaves of enterprise, drive, and initiative exist wherever it has been possible to escape the deadening hand of government control.

In any event, economic and social progress do not depend on the attributes or behavior of the masses. In every country a tiny minority sets the pace, determines the course of events. In the countries that have developed most rapidly and successfully, a minority of enterprising and risk-taking individuals have forged ahead, created opportunities for imitators to follow, have enabled the majority to increase their productivity. The characteristics of the Indians that so many outside observers deplore reflect rather than cause the lack of progress. Sloth and lack of enterprise flourish when hard work and the taking of risks are not rewarded. A fatalistic philosophy is an accommodation to stagnation. India has no shortage of people with the qualities that could spark and fuel the same kind of economic development that Japan experienced after 1867, or even that Germany and Japan did after World War II. Indeed, the real tragedy of India is that it remains a subcontinent teeming with desperately poor people when it could, we believe, be a flourishing, vigorous, increasingly prosperous and free society.

We recently came across a fascinating example of how an economic system can affect the qualities of people. Chinese refugees who streamed into Hong Kong after the communists gained power sparked its remarkable economic development and gained a deserved reputation for initiative, enterprise, thrift, and hard work. The recent liberalization of emigration from Red China has produced a new stream of immigrants—from the same racial stock, with the same fundamental cultural traditions, but raised and formed by thirty years of communist rule. We hear from several firms that hired some of these refugees that they are very different from the earlier Chinese entrants into Hong Kong. The new immigrants show little initiative and want to be told precisely what to do. They are indolent and uncooperative. No doubt a few years in Hong Kong's free market will change all that.

What then accounts for the different experiences of Japan from 1867 to 1897 and of India from 1947 to date? We believe that the explanation is the same as for the difference between West and East Germany, Israel and Egypt, Taiwan and Red China. Japan relied primarily on voluntary cooperation and free markets—on the model of the Britain of its time. India relied on central economic planning—on the model of the Britain of its time.

The Meiji government did intervene in many ways and played a key role in the process of development. It sent many Japanese abroad for technical training. It imported foreign experts. It established pilot plants in many industries and gave numerous subsidies to others. But at no time did it try to control the total amount or direction of investment or the structure of output. The state maintained a large interest only in shipbuilding and iron and steel industries that it thought necessary for military power. It retained these industries because they were not attractive to private enterprise and required heavy government subsidies. These subsidies were a drain on Japanese resources. They impeded rather than stimulated Japanese economic progress. Finally, an international treaty prohibited Japan during the first three decades after the Meiji Restoration from imposing tariffs higher than 5 percent. This restriction proved an unmitigated boon to Japan, though it was resented at the time, and tariffs were raised after the treaty prohibitions expired.

India is following a very different policy. Its leaders regard capitalism as synonymous with imperialism, to be avoided at all costs. They embarked on a series of Russian-type five-year plans that outlined detailed programs of investment. Some areas of production are reserved to government; in others private firms are permitted to operate, but only in conformity with The Plan. Tariffs and quotas control imports, subsidies control exports. Self-sufficiency is the ideal. Needless to say, these measures produce shortages of foreign exchange. These are met by detailed and extensive foreign exchange control—a major source both of inefficiency and of special privilege. Wages and prices are controlled. A government permit is required to build a factory or to make any other investment. Taxes are ubiquitous, highly graduated on paper, evaded in practice. Smuggling, black markets, illegal transactions of all kinds are every bit as ubiquitous as taxes, undermining all respect for law, yet performing a valuable social service by offsetting to some extent the rigidity of central planning and making it possible for urgent needs to be satisfied.

Reliance on the market in Japan released hidden and unsuspected resources of energy and ingenuity. It prevented vested interests from blocking change. It forced development to conform to the harsh test of efficiency. Reliance on government controls in India frustrates initiative or diverts it into wasteful channels.

It protects vested interests from the forces of change. It substitutes bureaucratic approval for market efficiency as the criterion of survival.

The experience in the two countries with homemade and factory-made textiles serves to illustrate the difference in policy. Both Japan in 1867 and India in 1947 had extensive production of textiles in the home. In Japan foreign competition did not have much effect on the home production of silk, perhaps because of Japan's advantage in raw silk reinforced by the failure of the European crop, but it all but wiped out the home spinning of cotton and later the hand-loom weaving of cotton cloth. A Japanese factory textile industry developed. At first, it manufactured only the coarsest and lowest-grade fabrics, but then moved to higher and higher grades and ultimately became a major export industry.

In India hand-loom weaving was subsidized and guaranteed a market, allegedly to ease the transition to factory production. Factory production is growing gradually but has been deliberately held back to protect the hand-loom industry. Protection has meant expansion. The number of hand looms roughly doubled from 1948 to 1978. Today, in thousands of villages throughout India, the sound of hand looms can be heard from early morning to late at night. There is nothing wrong with a hand-loom industry, provided it can compete on even terms with other industries. In Japan a prosperous, though extremely small, hand-loom industry still exists. It weaves luxury silk and other fabrics. In India the hand-loom industry prospers because it is subsidized by the government. Taxes are, in effect, imposed on people who are no better off than the ones who operate the looms in order to pay them a higher income than they could earn in a free market.

Early in the nineteenth century Great Britain faced precisely the same problem that Japan did a few decades later and India did more than a century later. The power loom threatened to destroy a prosperous hand-loom weaving industry. A royal commission was appointed to investigate the industry. It considered explicitly the policy followed by India: subsidizing hand-loom weaving and guaranteeing the industry a market. It rejected that policy out of hand on the ground that it would only make the basic problem, an excess of hand-loom weavers, worse—precisely what happened in India. Britain adopted the same solution as Japan—the temporarily harsh but ultimately beneficent policy of letting market forces work.[2]

The contrasting experiences of India and Japan are interesting because they bring out so clearly not only the different results of the two methods of organization but also the lack of relation between objectives pursued and policies adopted. The objectives of the new Meiji rulers—who were dedicated to strengthening the power and glory of their country and who attached little value to individual freedom—were more in tune with the Indian policies than with those they themselves adopted. The objectives of the new Indian leaders—who were ardently devoted to individual freedom—were more in tune with the Japanese policies than with those they themselves adopted.

CONTROLS AND FREEDOM

Though the United States has not adopted central economic planning, we have gone very far in the past fifty years in expanding the role of government in the economy. That intervention has been costly in economic terms. The limitations imposed on our economic freedom threaten to bring two centuries of economic progress to an end. Intervention has also been costly in political terms. It has greatly limited our human freedom.

The United States remains a predominantly free country—one of the freest major countries in the world. However, in the words of Abraham Lincoln's famous "House Divided" speech, "A house divided against itself cannot stand. . . . I do not expect the house to fall, but I do expect it will cease to be divided. It will become all one thing or all the other." He was talking about human slavery. His prophetic words apply equally to government intervention into the economy. Were it to go much further, our divided house would fall on the collectivist side. Fortunately, evidence grows that the public is recognizing the danger and is determined to stop and reverse the trend toward ever bigger government.

All of us are affected by the status quo. We tend to take for granted the situation as it is, to regard it as the natural state of affairs, especially when it has been shaped by a series of small gradual changes. It is hard to appreciate how great the cumulative effect has been. It takes an effort of the imagination to get outside the existing situation and view it with fresh eyes. The effort is well worth making. The result is likely to come as a surprise, not to say a shock.

Economic Freedom

An essential part of economic freedom is freedom to choose how to use our income: how much to spend on ourselves and on what items; how much to save and in what form; how much to give away and to whom. Currently, more than 40 percent of our income is disposed of on our behalf by government at federal, state, and local levels combined. One of us once suggested a new national holiday, "Personal Independence Day—that day in the year when we stop working to pay the expenses of government . . . and start working to pay for the items we severally and individually choose in light of our own needs and desires." [3] In 1929 that holiday would have come on Abraham Lincoln's birthday, February 12; today it would come about May 30; if present trends were to continue, it would coincide with the other Independence Day, July 4, around 1988.

Of course, we have something to say about how much of our income is spent on our behalf by government. We participate in the political process that has resulted in government's spending an amount equal to more than 40 percent of our income. Majority rule is a necessary and desirable expedient. It is, however, very different from the kind of freedom you have when you shop at a supermarket. When you enter the voting booth once a year, you almost always vote for a package rather than for specific items. If you are in the majority, you will at best get both the items you favored and the ones you opposed but regarded as on balance less important. Generally, you end up with something different from what you thought you voted for. If you are in the minority, you must conform to the majority vote and wait for your turn to come. When you vote daily in the supermarket, you get precisely what you voted for, and so does everyone else. The ballot box produces conformity without unanimity; the marketplace, unanimity without conformity. That is why it is desirable to use the ballot box, so far as possible, only for those decisions where conformity is essential.

As consumers, we are not even free to choose how to spend the part of our income that is left after taxes. We are not free to buy cyclamates or laetrile, and soon, perhaps, saccharin. Our physician is not free to prescribe many drugs for us that he may regard as the most effective for our ailments, even though the drugs may be widely available abroad. We are not free to buy an automobile without seat belts, though, for the time being, we are still free to choose whether or not to buckle up.

Another essential part of economic freedom is freedom to use the resources we possess in accordance with our own values— freedom to enter any occupation, engage in any business enterprise, buy from and sell to anyone else, so long as we do so on a strictly voluntary basis and do not resort to force in order to coerce others.

Today you are not free to offer your services as a lawyer, a physician, a dentist, a plumber, a barber, a mortician, or engage in a host of other occupations, without first getting a permit or license from a government official. You are not free to work overtime at terms mutually agreeable to you and your employer, unless the terms conform to rules and regulations laid down by a government official.

You are not free to set up a bank, go into the taxicab business, or the business of selling electricity or telephone service, or running a railroad, busline, or airline, without first receiving permission from a government official.

You are not free to raise funds on the capital markets unless you fill out the numerous pages of forms the SEC requires and unless you satisfy the SEC that the prospectus you propose to issue presents such a bleak picture of your prospects that no investor in his right mind would invest in your project if he took the prospectus literally. And getting SEC approval may cost upwards of $100,000—which certainly discourages the small firms our government professes to help.

Freedom to own property is another essential part of economic freedom. And we do have widespread property ownership. Well over half of us own the homes we live in. When it comes to machines, factories, and similar means of production, the situation is very different. We refer to ourselves as a free private enterprise society, as a capitalist society. Yet in terms of the ownership of corporate enterprise, we are about 46 percent socialist. Owning 1 percent of a corporation means that you are entitled to receive 1 percent of its profits and must share 1 percent of its losses up to the full value of your stock. The 1979 federal corporate income tax is 46 percent on all income over $100,000 (reduced from 48 percent in prior years). The federal government is entitled to 46 cents out of every dollar of profit, and it shares 46 cents out of every dollar of losses (provided there are some earlier profits to offset those losses). The federal government owns 46 percent of every corporation—though not in a form that entitles it to vote directly on corporate affairs.

It would take a book much longer than this one even to list in full all the restrictions on our economic freedom, let alone describe them in detail. These examples are intended simply to suggest how pervasive such restrictions have become.

Human Freedom

Restrictions on economic freedom inevitably affect freedom in general, even such areas as freedom of speech and press.

Consider the following excerpts from a 1977 letter from Lee Grace, then executive vice-president of an oil and gas association. This is what he wrote with respect to energy legislation:

> As you know, the real issue more so than the price per thousand cubic feet is the continuation of the First Amendment of the Constitution, the guarantee of freedom of speech. With increasing regulation, as big brother looks closer over our shoulder, we grow timid against speaking out for truth and our beliefs against falsehoods and wrong doings. Fear of IRS audits, bureaucratic strangulation or government harassment is a powerful weapon against freedom of speech.
> In the October 31 [1977] edition of the U.S. News & World Report, the Washington Whispers section noted that, "Oil industry officials claim that they have received this ultimatum from Energy Secretary

James Schlesinger: 'Support the Administration's proposed tax on crude oil—or else face tougher regulation and a possible drive to break up the oil companies.' "

His judgment is amply confirmed by the public behavior of oil officials. Tongue-lashed by Senator Henry Jackson for earning "obscene profits," not a single member of a group of oil industry executives answered back, or even left the room and refused to submit to further personal abuse. Oil company executives, who in private express strong opposition to the present complex structure of federal controls under which they operate or to the major extension of government intervention proposed by President Carter, make bland public statements approving the objectives of the controls.

Few businessmen regard President Carter's so-called voluntary wage and price controls as a desirable or effective way to combat inflation. Yet one businessman after another, one business organization after another, has paid lip service to the program, said nice things about it, and promised to cooperate. Only a few, like Donald Rumsfeld, former congressman, White House official, and Cabinet member, had the courage to denounce it publicly. They were joined by George Meany, the crusty octogenarian former head of the AFL-CIO.

It is entirely appropriate that people should bear a cost—if only of unpopularity and criticism—for speaking freely. However, the cost should be reasonable and not disproportionate. There should not be, in the words of a famous Supreme Court decision, "a chilling effect" on free speech. Yet there is little doubt that currently there is such an effect on business executives.

The "chilling effect" is not restricted to business executives. It affects all of us. We know most intimately the academic community. Many of our colleagues in economics and the natural science departments receive grants from the National Science Foundation; in the humanities, from the National Foundation for the Humanities; all those who teach in state universities get their salaries partly from the state legislatures. We believe that the National Science Foundation, the National Foundation for the Humanities, and tax subsidies to higher education are all undesirable and should be terminated. That is undoubtedly a minority view in the academic community, but the minority is much larger than anyone would gather from public statements to that effect.

The press is highly dependent on government—not only as a major source of news but in numerous other day-to-day operating matters. Consider a striking example from Great Britain. The London *Times*, a great newspaper, was prevented from publishing one day several years ago by one of its unions because of a story that it was planning to publish about the union's attempt to influence the content of the paper. Subsequently, labor disputes closed down the paper entirely. The unions in question are able to exercise this power because they have been granted special immunities by government. A national Union of Journalists in Britain is pushing for a closed shop of journalists and threatening to boycott papers that employ nonmembers of the union. All this in the country that was the source of so many of our liberties.

With respect to religious freedom, Amish farmers in the United States have had their houses and other property seized because they refused, on religious grounds, to pay Social Security taxes—and also to accept Social Security benefits. Church schools have had their students cited as truants in violation of compulsory attendance laws because their teachers did not have the requisite slips of paper certifying to their having satisfied state requirements.

Although these examples only scratch the surface, they illustrate the fundamental proposition that freedom is one whole, that anything that reduces freedom in one part of our lives is likely to affect freedom in the other parts.

Freedom cannot be absolute. We do live in an interdependent society. Some restrictions on our freedom are necessary to avoid other, still worse, restrictions. However, we have gone far beyond that point. The urgent need today is to eliminate restrictions, not add to them.

CHAPTER 3

The Anatomy
of Crisis

The depression that started in mid-1929 was a catastrophe of unprecedented dimensions for the United States. The dollar income of the nation was cut in half before the economy hit bottom in 1933. Total output fell by a third, and unemployment reached the unprecedented level of 25 percent of the work force. The depression was no less a catastrophe for the rest of the world. As it spread to other countries, it brought lower output, higher unemployment, hunger, and misery everywhere. In Germany the depression helped Adolf Hitler rise to power, paving the way for World War II. In Japan it strengthened the military clique that was dedicated to creating a Greater East Asia coprosperity sphere. In China it led to monetary changes that accelerated the final hyperinflation that sealed the doom of the Chiang Kai-shek regime and brought the communists to power.

In the realm of ideas, the depression persuaded the public that capitalism was an unstable system destined to suffer ever more serious crises. The public was converted to views that had already gained increasing acceptance among the intellectuals: government had to play a more active role; it had to intervene to offset the instability generated by unregulated private enterprise; it had to serve as a balance wheel to promote stability and assure security. The change in the public's perception of the proper role of private enterprise on the one hand and of the government on the other proved a major catalyst for the rapid growth of government, and particularly central government, from that day to this.

The depression also produced a far-reaching change in professional economic opinion. The economic collapse shattered the long-held belief, which had been strengthened during the 1920s, that monetary policy was a potent instrument for promoting economic stability. Opinion shifted almost to the opposite extreme, that "money does not matter." John Maynard Keynes, one of the great economists of the twentieth century, offered an alternative theory. The Keynesian revolution not only captured the economics profession, but also provided both an appealing justification and a prescription for extensive government intervention.

The shift in opinion of both the public and the economics profession resulted from a misunderstanding of what had actually happened. We now know, as a few knew then, that the depression was not produced by a failure of private enterprise, but rather by a failure of government in an area in which the government had from the first been assigned responsibility—"To coin money, regulate the Value thereof, and of foreign Coin," in the words of Section 8, Article 1, of the U.S. Constitution. Unfortunately, as we shall see in Chapter 9, government failure in managing money is not merely a historical curiosity but continues to be a present-day reality.

THE ORIGIN OF THE FEDERAL RESERVE SYSTEM

On Monday, October 21, 1907, some five months after the start of an economic recession, the Knickerbocker Trust Company, the third largest trust company in New York City, began to experience financial difficulties. The next day a "run" on the bank forced it to close (temporarily, as it turned out; it resumed business in March 1908). The closing of the Knickerbocker Trust precipitated runs on other trust companies in New York and then in other parts of the country—a banking "panic" was under way of a kind that had occurred every now and then during the nineteenth century.

Within a week, banks throughout the country reacted to the "panic" by "restriction of payments," i.e., they announced that they would no longer pay out currency on demand to depositors who wanted to withdraw their deposits. In some states the governor or attorney general took measures that gave legal sanction to the restriction of payments; in the remaining states the practice was simply tolerated and banks were permitted to stay open even though they were technically violating the state banking laws.

The restriction of payments cut short bank failures and ended the runs. But it imposed serious inconvenience on business. It led to a shortage of coin and currency, as well as to the private circulation of wooden nickels and other temporary substitutes for legal money. At the height of the shortage of currency, it took $104 of deposits to buy $100 of currency. Together, the panic and the restriction, both directly, through their effects on confidence and on the possibility of conducting business efficiently, and indirectly, by forcing a decline in the quantity of money, turned the recession into one of the most severe that the United States had experienced up to that time.

However, the severe phase of the recession was short-lived. Banks resumed payments in early 1908. A few months later, economic recovery began. The recession lasted only thirteen months in all, and its severe phase only about half that long.

This dramatic episode was largely responsible for the enactment of the Federal Reserve Act in 1913. It made some action in the monetary and banking area politically essential. During Theodore Roosevelt's Republican administration a National Monetary Commission was established that was headed by a prominent Republican senator, Nelson W. Aldrich. During Woodrow Wilson's Democratic administration, a prominent Democratic congressman, later senator, Carter Glass, rewrote and repackaged the commission's recommendations. The resulting Federal Reserve System has served as the key monetary authority of the country ever since.

What do the terms "run" and "panic" and "restriction of payments" really mean? Why did they have the far-reaching effects we have attributed to them? And how did the authors of the Federal Reserve Act propose to prevent similar episodes?

A run on a bank is an attempt by many of its depositors to "withdraw" their deposits in cash, all at the same time. The run arises from rumors or facts that lead depositors to fear that the bank is insolvent and will be unable to live up to its obligations. It represents an attempt by everyone to get "his" money out before it is all gone.

It is easy to see why a run would cause an insolvent bank to fail sooner than it otherwise might. But why should a run cause a responsible and solvent bank trouble? The answer is linked to one of the most misleading words in the English language—the word "deposit," when used to refer to a claim against a bank. If you "deposit" currency in a bank, it is tempting to suppose that the bank takes your greenbacks and "deposits" them in a bank vault for safekeeping until you ask for them. It does nothing of the kind. If it did, where would the bank get income to pay its expenses, let alone to pay interest on deposits? The bank may take a few of the greenbacks and put them in a vault as a "reserve." The rest it lends to someone else, charging the borrower interest, or uses to buy an interest-bearing security.

If, as is typically the case, you deposit not currency but checks on other banks, your bank does not even have currency in hand to deposit in a vault. It has only a claim on another bank for currency, which it typically will not exercise because other banks have matching claims on it. For every $100 of deposits, all the banks together have only a few dollars of cash in their vaults. We have a "fractional reserve banking system." That system works very well, so long as everyone is confident that he can always get cash for his deposits and therefore only tries to get cash when he really needs it. Usually, new deposits of cash roughly equal withdrawals, so that the small amount in reserve is sufficient to meet temporary discrepancies. However, if everyone tries to get cash at once, the situation is very different—a panic is likely to occur, just as it does when someone cries "fire" in a crowded theater and everyone rushes to get out.

One bank alone can meet a run by borrowing from other banks, or by asking its borrowers to repay their loans. The borrowers may be able to repay their loans by withdrawing cash from other banks. But if a bank run spreads widely, all banks together cannot meet the run in this way. There simply is not enough currency in bank vaults to satisfy the demands of all depositors. Moreover, any attempt to meet a widespread run by drawing down vault cash—unless it succeeds promptly in restoring confidence and ends the run so that the cash is redeposited —enforces a much larger reduction in deposits. On the average in 1907, the banks had only $12 of cash for every $100 of deposits. Every dollar of deposits converted into cash and transferred from the vaults of banks to the mattresses of depositors made necessary the reduction of deposits by an additional $7 if banks were to maintain the prior ratio of reserves to deposits. That is why a run that results in hoarding of cash by the public tends to reduce the total money supply. It is also why, if not stopped promptly, it causes such distress. Individual banks try to get cash to meet the demands of their depositors by pressing their borrowers to repay loans and by refusing to renew loans or to extend additional ones. Borrowers as a whole have nowhere to turn, so banks fail and businesses fail.

How can a panic be stopped once it is under way, or better yet, how can it be prevented from starting? One way to stop a panic is the method adopted in 1907: a concerted restriction of payments by the banks. Banks stayed open but they agreed with one another that they would not pay cash on demand to depositors. Instead, they operated through bookkeeping entries. They honored checks written by one of their own depositors to another by reducing the deposits recorded on their books to the credit of the one and increasing the deposits of the other. For checks written by their depositors to another bank's depositors, or by another bank's depositors to their depositors, they operated almost as usual "through the clearinghouse," that is, by offsetting the checks on other banks received as deposits against the checks on their own bank deposited in other banks. The one difference was that any differences between the amount they owed other banks and the amount other banks owed them was settled by a promise to pay instead of, as ordinarily, by the transfer of cash. Banks paid out some currency, not on demand, but to regular customers who needed it for payrolls and similar urgent purposes, and similarly, they received some currency from such regular customers. Under this system banks might and did still fail because they were "unsound" banks. They did not fail merely because they could not convert perfectly sound assets into cash. As time passed, panic subsided, confidence in banks was restored, and the banks could resume payment of cash on demand without starting a new series of runs. That is a rather drastic way to stop a panic but it worked.

Another way to stop a panic is to enable sound banks to convert their assets into cash rapidly, not at the expense of other banks but through the availability of additional cash—of an emergency printing press, as it were. This was the way embodied in the Federal Reserve Act. It was supposed to prevent even the temporary disruptions produced by the restriction of payments. The twelve regional banks established by the act, operating under the supervision of a Federal Reserve Board in Washington, were given the power to serve as "lenders of last resort" to the commercial banks. They could make such loans either in the form of currency—Federal Reserve Notes, which they had the power to print—or in the form of deposit credits on their books, which they had the power to create—the magic of the bookkeeper's pen. They were to serve as bankers' banks, as the U.S. counterpart of the Bank of England and other central banks.

Initially, it was expected that the Federal Reserve Banks would operate mostly by direct loans to banks, on the security of the banks' own assets, in particular, the promissory notes corresponding to loans by banks to businesses. In many such loans, the banks "discounted" the notes—that is, paid out less than the face amount, the discount representing the interest charged by the banks. The Federal Reserve in turn "rediscounted" the promissory notes, thereby charging the banks interest on the loans.

As time passed, "open market operations"—the purchase or sale of government bonds—rather than rediscounts became the main way in which the System added to or subtracted from the amount of money. When a Federal Reserve Bank buys a government bond, it pays for it either with Federal Reserve Notes that it has in its vaults or that it has freshly printed or, more typically, by adding on its books to the deposits of a commercial bank. The commercial bank may itself be the seller of the bond or it may be the bank in which the seller of the bond keeps his deposit account. The extra currency and deposits serve as reserves for the commercial banks, enabling them as a whole to expand their deposits by a multiple of the additional reserves, which is why currency plus deposits at Federal Reserve Banks are designated "high-powered money" or the "monetary base." When a Federal Reserve Bank sells a bond, the process is reversed. Reserves of commercial banks decline and they are led to contract. Until fairly recently the power of the Federal Reserve Banks to create currency and deposits was limited by the amount of gold held by the

System. That limit has now been removed so that today there is no effective limit except the discretion of the people in charge of the System.

After the Federal Reserve System failed in the early 1930s to do what it had been set up to do, an effective method of preventing a panic was finally adopted in 1934. The Federal Deposit Insurance Corporation was established to guarantee deposits against loss up to a maximum. The insurance gives depositors confidence that their deposits are safe. It thereby prevents the failure or financial difficulties of an unsound bank from creating runs on other banks. The people in the crowded theater are confident that it is really fireproof. Since 1934 there have been bank failures and some runs on individual banks. There have been no banking panics of the old style.

Guaranteeing deposits in order to prevent a panic had frequently been used earlier by the banks themselves in a more partial and less effective way. Time and again, when an individual bank was in financial trouble or was threatened by a run because of rumors of trouble, other banks banded together voluntarily to subscribe to a fund guaranteeing the deposits of the bank in trouble. That device prevented many putative panics and cut others short. It failed on other occasions either because a satisfactory agreement could not be reached or because confidence was not promptly restored. We shall examine a particularly dramatic and important case of such a failure later in this chapter.

THE EARLY YEARS OF THE RESERVE SYSTEM

The Federal Reserve System started to operate in late 1914, a few months after the outbreak of war in Europe. That war changed drastically the role and importance of the Federal Reserve System.

When the System was established, Britain was the center of the financial world. The world was said to be on a gold standard but it could equally well have been said to be on a sterling standard. The Federal Reserve System was envisioned primarily as a means of avoiding banking panics and facilitating commerce; secondarily, as the government's banker. It was taken for granted that it would operate within a world gold standard, reacting to external events but not shaping them.

By the end of the war the United States had replaced Britain as the center of the financial world. The world was effectively on a dollar standard and remained so even after a weakened version of the prewar gold standard was reestablished. The Federal Reserve System was no longer a minor body reacting passively to external events. It was a major independent force shaping the world monetary structure.

The war years demonstrated the power of the Federal Reserve System for both good and ill, especially after the United States entered the war. As in all previous (and subsequent) wars, the equivalent of the printing press was resorted to in order to finance war spending. But the System made it possible to do so in a more sophisticated and subtle manner than was possible earlier. The literal printing press was used to some extent, when the Federal Reserve Banks bought bonds from the U.S. Treasury and paid for them with Federal Reserve Notes that the Treasury could pay out to meet some of its expenses. But mostly the Fed paid for bonds it bought by crediting the Treasury with deposits at the Federal Reserve Banks. The Treasury paid for purchases with checks drawn on these deposits. When the recipients deposited the checks in their own banks and these banks in turn deposited them at a Federal Reserve Bank, the Treasury deposits at the Fed were transferred to the commercial banks, increasing their reserves. That increase enabled the commercial banking system to expand, largely—at the time—by buying government bonds themselves or making loans to their customers to enable them to buy government bonds. By this roundabout process, the Treasury got newly created money to pay for war expenses, but the increase in the quantity of money mostly took the form of increases in deposits at commercial banks, rather than of currency. The subtlety of the process whereby the quantity of money was increased did not prevent inflation, but it did smooth the operation and, by concealing what was actually happening, lessen or postpone the public's fears about inflation.

After the war the System continued to increase the quantity of money rapidly, thereby feeding the inflation. At this stage, however, the additional money was being used not to pay for the government's expenses but to finance private business activities. A third of our total wartime inflation occurred after the end not only of the war but also of government deficits to pay for the war.

Belatedly, the System discovered its mistake. It then reacted sharply, plunging the country into the sharp but short depression of 1920–21.

The high tide of the System was undoubtedly the rest of the twenties. During those few years it did serve as an effective balance wheel, increasing the rate of monetary growth when the economy showed signs of faltering, and reducing the rate of monetary growth when the economy started expanding more rapidly. It did not prevent fluctuations in the economy but it did contribute to keeping them mild. Moreover, it was sufficiently evenhanded so that it avoided inflation. The result of the stable monetary and economic climate was rapid economic growth. It was widely trumpeted that a new era had arrived, that the business cycle was dead, dispatched by a vigilant Federal Reserve System.

Much of the success during the twenties can be credited to Benjamin Strong, a New York banker who was the first head of the Federal Reserve Bank of New York and remained its head until his untimely death in 1928. Until he died, the New York Bank was the prime mover in Federal Reserve policy both at home and abroad, and Benjamin Strong was unquestionably the dominant figure. He was a remarkable man, described by a member of the Federal Reserve Board as "a genius—a Hamilton among bankers." More than any other individual in the System, he had the confidence and backing of other financial leaders inside and outside the System, the personal force to make his views prevail, and the courage to act upon them.

Strong's death unleashed a struggle for power within the System that was fated to have far-reaching consequences. As Strong's biographer puts it, "Strong's death left the System with no center of enterprising and acceptable leadership. The Federal Reserve Board [in Washington] was determined that the New York Bank should no longer play that role. But the Board itself could not play the role in an enterprising way. It was still weak and divided. . . . Moreover, most of the other Reserve Banks, as well as that in New York, were reluctant to follow the leadership of the Board. . . . Thus it was easy for the System to slide into indecision and deadlock." [1]

This struggle for power proved to be—as no one could have

foreseen at the time—the first step in a greatly speeded-up transfer of power from the private market to government, and from local and state government to Washington.

THE ONSET OF DEPRESSION

The popular view is that the depression started on Black Thursday, October 24, 1929, when the New York stock market collapsed. After several intermediate ups and downs, the market ended up in 1933 at about one-sixth the dizzying level of 1929. The stock market crash was important, but it was not the beginning of the depression. Business activity reached its peak in August 1929, two months before the stock market crashed, and had already fallen appreciably by then. The crash reflected the growing economic difficulties plus the puncturing of an unsustainable speculative bubble. Of course, once the crash occurred, it spread uncertainty among businessmen and others who had been bemused by dazzling hopes of a new era. It dampened the willingness of both consumers and business entrepreneurs to spend and enhanced their desire to increase their liquid reserves for emergencies.

These depressing effects of the stock market crash were strongly reinforced by the subsequent behavior of the Federal Reserve System. At the time of the crash, the New York Federal Reserve Bank, almost by conditioned reflex instilled during the Strong era, immediately acted on its own to cushion the shock by purchasing government securities, thereby adding to bank reserves. That enabled commercial banks to cushion the shock by providing additional loans to stock market firms and purchasing securities from them and others affected adversely by the crash. But Strong *was* dead, and the Board wanted to establish its leadership. It moved rapidly to impose its discipline on New York, and New York yielded. Thereafter the System acted very differently than it had during earlier economic recessions in the 1920s. Instead of actively expanding the money supply by more than the usual amount to offset the contraction, the System allowed the quantity of money to decline slowly throughout 1930. Compared to the decline of roughly one-third in the quantity of money from late 1930 to early 1933, the decline in the quantity of money up to October 1930 seems mild—a mere 2.6 percent. However, by comparison with past episodes, it was sizable. Indeed, it was a larger decline than had occurred during or preceding all but a few of the earlier recessions.

The combined effect of the aftermath of the stock market crash and the slow decline in the quantity of money during 1930 was a rather severe recession. Even if the recession had come to an end in late 1930 or early 1931, as it might well have done if a monetary collapse had not occurred, it would have ranked as one of the most severe recessions on record.

BANKING CRISES

But the worst was yet to come. Until the autumn of 1930 the contraction, though severe, was not marred by banking difficulties or runs on banks. The character of the recession changed drastically when a series of bank failures in the Middle West and South undermined confidence in banks and led to widespread attempts to convert deposits into currency.

The contagion finally spread to New York, the financial center of the country. The critical date is December 11, 1930, when the Bank of United States closed its doors. It was the largest commercial bank that had ever failed up to that time in U.S. history. In addition, although it was an ordinary commercial bank, its name led many at home and abroad to regard it as an official bank. Its failure was therefore a particularly serious blow to confidence.

It was something of an accident that the Bank of United States played such a key role. Given the decentralized structure of the U.S. banking system plus the policy that the Federal Reserve System was following of letting the money stock decline and not responding vigorously to bank failures, the stream of minor failures would sooner or later have produced runs on other major banks. If the Bank of United States had not failed when it did, the failure of another major bank would have been the pebble that started the avalanche. It was also an accident that the Bank of United States itself failed. It was a sound bank. Though liquidated during the worst years of the depression, it ended up paying off depositors 83.5 cents on the dollar. There is little doubt that if it had been able to weather the immediate crisis, no depositor would have lost a cent.

When rumors started to spread about the Bank of United States, the New York State Superintendent of Banks, the Federal Reserve Bank of New York, and the New York Clearing House Association of Banks tried to devise plans to save the bank, through providing a guarantee fund or merging it with other banks. This had been the standard pattern in earlier panics. Until two days before the bank closed, these efforts seemed assured of success.

The plan failed, however, primarily because of the particular character of the Bank of United States plus the prejudices of the banking community. The name itself, because it appealed to immigrants, was resented by other banks. Far more important, the bank was owned and managed by Jews and served mostly the Jewish community. It was one of a handful of Jewish-owned banks in an industry that, more than almost any other, has been the preserve of the well-born and well-placed. By no accident, the planned rescue involved merging the Bank of United States with the only other major bank in New York City that was largely owned and run by Jews, plus two much smaller Jewish-owned banks.

The plan failed because the New York Clearing House at the last moment withdrew from the proposed arrangement—purportedly in large part because of the anti-Semitism of some of the leading members of the banking community. At the final meeting of the bankers, Joseph A. Broderick, then the New York State Superintendent of Banks, tried but failed to get them to go along. "I said," he later testified at a court trial,

> it [the Bank of United States] had thousands of borrowers, that it financed small merchants, especially Jewish merchants, and that its closing might and probably would result in widespread bankruptcy among those it served. I warned that its closing would result in the closing of at least 10 other banks in the city and that it might even affect the savings banks. The influence of the closing might even extend outside the city, I told them.

I reminded them that only two or three weeks before they had rescued two of the largest private bankers of the city and had willingly put up the money needed. I recalled that only seven or eight years before that they had come to the aid of one of the biggest trust companies in New York, putting up many times the sum needed to save the Bank of United States but only after some of their heads had been knocked together.

I asked them if their decision to drop the plan was still final. They told me it was. Then I warned them that they were making the most colossal mistake in the banking history of New York.[2]

The closing of the Bank of United States was tragic for its owners and depositors. Two of the owners were tried, convicted, and served prison sentences for what everybody agreed were technical infractions of the law. The depositors had even that part of their funds that they finally recovered tied up for years. For the country as a whole the effects were more far-reaching. Depositors all over the country, frightened about the safety of their deposits, added to the sporadic runs that had started earlier. Banks failed by the droves—352 banks in the month of December 1930 alone.

Had the Federal Reserve System never been established, and had a similar series of runs started, there is little doubt that the same measures would have been taken as in 1907—a restriction of payments. That would have been more drastic than what actually occurred in the final months of 1930. However, by preventing the draining of reserves from good banks, restriction would almost certainly have prevented the subsequent series of bank failures in 1931, 1932, and 1933, just as restriction in 1907 quickly ended bank failures then. Indeed, the Bank of United States itself might have been able to reopen, as the Knickerbocker Trust Company had in 1908. The panic over, confidence restored, economic recovery would very likely have begun in early 1931, just as it had in early 1908.

The existence of the Reserve System prevented this drastic therapeutic measure: directly, by reducing the concern of the stronger banks, who, mistakenly as it turned out, were confident that borrowing from the System offered them a reliable escape mechanism in case of difficulty; indirectly, by lulling the community as a whole, and the banking system in particular, into the belief that such drastic measures were no longer necessary now that the System was there to take care of such matters.

The System could have provided a far better solution by engaging in large-scale open market purchases of government bonds. That would have provided banks with additional cash to meet the demands of their depositors. That would have ended—or at least sharply reduced—the stream of bank failures and have prevented the public's attempted conversion of deposits into currency from reducing the quantity of money. Unfortunately, the Fed's actions were hesitant and small. In the main, it stood idly by and let the crisis take its course—a pattern of behavior that was to be repeated again and again during the next two years.

It was repeated in the spring of 1931, when a second banking crisis developed. An even more perverse policy was followed in September 1931, when Britain abandoned the gold standard. The Fed reacted—after two years of severe depression—by *raising* the rate of interest (the discount rate) that it charged banks for loans more sharply than ever before in its history. It took this action to avert a drain on its gold reserves by foreign holders of dollars that it feared would be set off by Britain's abandonment of the gold standard. The effect domestically, however, was highly deflationary—putting further pressure on both commercial banks and business enterprises. The Fed could, by open market purchases of government securities, have offset this sharp monetary blow that it gave to a struggling economy, but it did not do so.

In 1932, under strong pressure from Congress, the Fed finally undertook large-scale open market purchases. The favorable effects were just starting to be felt when Congress adjourned—and the Fed promptly terminated its program.

The final episode in this sorry tale was the banking panic of 1933, once again initiated by a series of bank failures. It was intensified by the interregnum between Herbert Hoover and Franklin D. Roosevelt, who was elected on November 8, 1932, but not inaugurated until March 4, 1933. Herbert Hoover was unwilling to take drastic measures without the cooperation of the President-elect, and FDR was unwilling to assume any responsibility until he was inaugurated.

As panic spread in the New York financial community, the System itself panicked. The head of the New York Federal Reserve Bank tried unsuccessfully to persuade President Hoover to declare a national banking holiday on Hoover's last day in office.

He then joined with the New York Clearing House banks and the State Superintendent of Banks to persuade Governor Lehman of New York* to declare a state banking holiday effective on March 4, 1933, the day of FDR's inauguration. The Federal Reserve Bank closed along with the commercial banks. Similar actions were taken by other governors. A nationwide holiday was finally proclaimed by President Roosevelt on March 6.

The central banking system, set up primarily to render unnecessary the restriction of payments by commercial banks, itself joined the commercial banks in a more widespread, complete, and economically disturbing restriction of payments than had ever been experienced in the history of the country. One can certainly sympathize with Hoover's comment in his memoirs: "I concluded it [the Reserve Board] was indeed a weak reed for a nation to lean on in time of trouble." [3]

At the peak of business in mid-1929, nearly 25,000 commercial banks were in operation in the United States. By early 1933 the number had shrunk to 18,000. When the banking holiday was ended by President Roosevelt ten days after it began, fewer than 12,000 banks were permitted to open, and only 3,000 additional banks were later permitted to do so. All in all, therefore, roughly 10,000 out of 25,000 banks disappeared during those four years —through failure, merger, or liquidation.

The total stock of money showed an equally drastic decline. For every $3 of deposits and currency in the hands of the public in 1929, less than $2 remained in 1933—a monetary collapse without precedent.

FACTS AND INTERPRETATION

These facts are not in question today—though it should be stressed that they were not known or available to most contemporary observers, including John Maynard Keynes. But they are susceptible of different interpretations. Was the monetary collapse a cause of the economic collapse or a result? Could the System have prevented the monetary collapse? Or did it happen in spite of the best efforts of the Fed—as many observers at the time concluded? Did the depression start in the United States and spread abroad? Or did forces emanating from abroad convert what might have been a fairly mild recession in the United States into a severe one?

Cause or Effect

The System itself expressed no doubt about its role. So great is the capacity for self-justification that the Federal Reserve Board could say in its *Annual Report* for 1933, "The ability of the Federal Reserve Banks to meet enormous demands for currency during the crisis demonstrated the effectiveness of the country's currency system under the Federal Reserve Act. . . . It is difficult to say what the course of the depression would have been had the Federal Reserve System not pursued a policy of liberal open market purchases." [4]

The monetary collapse was both a cause and an effect of the economic collapse. It originated in large measure from Federal Reserve policy, and it unquestionably made the economic collapse far worse than it would otherwise have been. However, the economic collapse, once it started, made the monetary collapse worse. Banks loans that might have been "good" loans in a milder recession became "bad" loans in the severe economic collapse. Defaults on loans weakened the lending banks, which added to the temptation for depositors to start a run on them. Business failures, declining output, growing unemployment—all fostered uncertainty and fear. The desire to convert assets into their most liquid form, money, and into the safest kind of money, currency, became widespread. "Feedback" is a pervasive feature of an economic system.

The evidence by now is all but conclusive that the System not only had a legislative mandate to prevent the monetary collapse, but could have done so if it had used wisely the powers that had been granted to it in the Federal Reserve Act. Defenders of the System have offered a series of excuses. None has withstood careful examination. None is a valid justification for the failure of the System to perform the task for which its founders had established it. The System not only had the power to prevent the monetary collapse, it also knew how to use that power. In 1929, 1930, 1931, the New York Federal Reserve Bank repeatedly urged the

System to engage in large-scale open market purchases, the key action the System should have taken but did not. New York was overruled not because its proposals were demonstrated to be misdirected or not feasible but because of the struggle for power within the System, which made both other Federal Reserve Banks and the Board in Washington unwilling to accept New York's leadership. The alternative proved to be confused and indecisive leadership by the Board. Knowledgeable voices outside the System also called for the correct action. An Illinois congressman, A. J. Sabath, said on the floor of the House, "I insist it is within the power of the Federal Reserve Board to relieve the financial and commercial distress." Some academic critics—including Karl Bopp, who later became the head of the Federal Reserve Bank of Philadelphia—expressed similar views. At the Federal Reserve meeting at which the 1932 open market purchases were approved, under direct pressure from the Congress, Ogden L. Mills, then Secretary of the Treasury and an ex officio member of the Board, stated, in explaining his vote for the action, "For a great central banking system to stand by with a 70% gold reserve without taking active steps in such a situation was almost inconceivable and almost unforgivable." Yet that was precisely how the System had behaved for the two prior years and was to resume behaving as soon as Congress adjourned a few months later, as well as during the climactic final banking crisis of March 1933.[5]

Where the Depression Started

The decisive evidence that the depression spread from the United States to the rest of the world, rather than the other way around, comes from the movement of gold. In 1929 the United States was on a gold standard in the sense that there was an official price of gold ($20.67 per fine ounce) at which the U.S. government would buy or sell gold on demand. Most other major countries were on a so-called gold-exchange standard, under which they, too, specified an official price for gold in terms of their own currencies. That official price of gold in their currency divided by the U.S. official price gave an official exchange rate, that is, the price of their currency in terms of the dollar. They might or might not buy and sell gold freely at the official price, but they committed themselves to keep the exchange rate fixed at the level determined by the two official prices of gold by buying and selling dollars on demand at that exchange rate. Under such a system, if United States residents, or others who had dollars, spent (or lent or gave) abroad more dollars than the recipients of those dollars wanted to spend (or lend or give) in the United States, the recipients would demand gold for the difference. Gold would go from the United States to foreign countries. If the balance was in the opposite direction, so that holders of foreign currencies wanted to spend (or lend or give) more dollars in the United States than holders of dollars wanted to convert into foreign currencies to spend (or lend or give) abroad, they would get the extra dollars by buying them from their central banks at the official exchange rates. The central banks, in turn, would get the extra dollars by sending gold to the United States. (In practice, of course, most of these transfers did not involve the literal shipping of gold across the oceans. Much of the gold owned by foreign central banks was stored in the vaults of the New York Federal Reserve Bank, "earmarked" for the country that owned it. The transfer was made by changing the labels on the containers holding the gold bars in the deep basements under the bank building at 33 Liberty Street in the Wall Street area.)

If the depression had originated abroad while the U.S. economy continued, for a time, to boom, the deteriorating economic conditions abroad would have reduced U.S. exports and, by lowering the cost of foreign goods, encouraged U.S. imports. The result would have been an attempt to spend (or lend or give) more dollars abroad than recipients wanted to use in the United States and an outflow of gold from the United States. The outflow of gold would have reduced the Federal Reserve System's gold reserves. And that would, in turn, have induced the System to reduce the quantity of money. That is how a system of fixed exchange rates transmits deflationary (or inflationary) pressure from one country to another. If this had been the course of events, the Federal Reserve could correctly have claimed that its actions were a response to pressures coming from abroad.

Conversely, if the depression originated in the United States, an early effect would be a decline in the number of U.S. dollars that their holders wanted to use abroad and an increase in the number of dollars that others wanted to use in the United States. That would have produced an inflow of gold into the United States. That, in turn, would bring pressure on foreign countries to reduce their quantity of money and would be the way the U.S. deflation would be transmitted to them.

The facts are clear. The U.S. gold stock *rose* from August 1929 to August 1931, the first two years of the contraction—clinching evidence that the United States was in the van of the movement. Had the Federal Reserve System followed the rules of the gold standard, it should have reacted to the inflow of gold by increasing the quantity of money. Instead, it actually let the quantity of money decline.

Once the depression was under way and had been transmitted to other countries, there was, of course, a reflex influence on the United States—another example of the feedback that is so ubiquitous in any complex economy. The country in the vanguard of an international movement need not stay there. France had accumulated a large stock of gold as a result of returning to the gold standard in 1928 at an exchange rate that undervalued the franc. It therefore had much leeway and could have resisted the deflationary pressure coming from the United States. Instead, France followed even more deflationary policies than the United States and not only began to add to its large gold stock but also, after late 1931, to drain gold from the United States. Its dubious reward for such leadership was that, although the U.S. economy hit bottom when it suspended gold payments in March 1933, the French economy did not hit bottom until April 1935.

Effect on the Reserve System

One ironic result of the perverse monetary policy of the Federal Reserve Board, despite the good advice of the New York Federal Reserve Bank, was a complete victory for the Board against both New York and the other Federal Reserve Banks in the struggle for power. The myth that private enterprise, including the private banking system, had failed, and that government needed more power to counteract the alleged inherent instability of the free market, meant that the System's failure produced a political environment favorable to giving the Board greater control over the regional banks. One symbol of the change was the transfer of the Federal Reserve Board from modest offices in the U.S. Treasury Building to a magnificent Greek temple of its own on Constitution Avenue (since supplemented by a massive additional structure).

The final seal on the shift of power was a change in the name of the Board and in the title of the head officers of the regional banks. In central bank circles the prestigious title is Governor, not President. From 1913 to 1935, the head of a regional bank was designated "Governor"; the central Washington body was called "The Federal Reserve Board"; only the chairman of the Board was designated "Governor"; the remaining members were simply "members of the Federal Reserve Board." The Banking Act of 1935 changed all that. The heads of the regional banks were designated "Presidents" instead of "Governors"; and the compact "Federal Reserve Board" was replaced by the cumbrous "Board of Governors of the Federal Reserve System," solely in order that each of the members of the Board could be designated a "Governor."

Unfortunately, the increase in power, prestige, and trappings of office has not been accompanied by a corresponding improvement in performance. Since 1935 the System has presided over— and greatly contributed to—a major recession in 1937–38, a wartime and immediate postwar inflation, and a roller coaster economy since, with alternate rises and falls in inflation and decreases and increases in unemployment. Each inflationary peak and each temporary inflationary trough has been at a higher and higher level, and the average level of unemployment has gradually increased. The System has not made the same mistake that it made in 1929–33—of permitting or fostering a monetary collapse—but it has made the opposite mistake, of fostering an unduly rapid growth in the quantity of money and so promoting inflation. In addition, it has continued, by swinging from one extreme to another, to produce not only booms but also recessions, some mild, some sharp.

In one respect the System has remained completely consistent throughout. It blames all problems on external influences beyond its control and takes credit for any and all favorable occurrences. It thereby continues to promote the myth that the private economy is unstable, while its behavior continues to document the reality that government is today the major source of economic instability.

CHAPTER 4

Cradle
to Grave

The presidential election of 1932 was a political watershed for the United States. Herbert Hoover, seeking reelection on the Republican ticket, was saddled with a deep depression. Millions of people were unemployed. The standard image of the time was a breadline or an unemployed person selling apples on a street corner. Though the independent Federal Reserve System was to blame for the mistaken monetary policy that converted a recession into a catastrophic depression, the President, as the head of state, could not escape responsibility. The public had lost faith in the prevailing economic system. People were desperate. They wanted reassurance, a promise of a way out.

Franklin Delano Roosevelt, the charismatic governor of New York, was the Democratic candidate. He was a fresh face, exuding hope and optimism. True enough, he campaigned on the old principles. He promised if elected to cut waste in government and balance the budget, and berated Hoover for extravagance in government spending and for permitting government deficits to mount. At the same time, both before the election and during the interlude before his inauguration, he met regularly with a group of advisers at the Governor's Mansion in Albany—his "brain trust," as it was christened. They devised measures to be taken after his inauguration that grew into the "New Deal" FDR had pledged to the American people in accepting the Democratic nomination for President.

The election of 1932 was a watershed in narrowly political terms. In the seventy-two years from 1860 to 1932, Republicans held the presidency for fifty-six years, Democrats for sixteen. In the forty-eight years from 1932 to 1980, the tables were turned: Democrats held the presidency for thirty-two years, Republicans for sixteen.

The election was also a watershed in a more important sense;

it marked a major change in both the public's perception of the role of government and the actual role assigned to government. One simple set of statistics suggests the magnitude of the change. From the founding of the Republic to 1929, spending by governments at all levels, federal, state, and local, never exceeded 12 percent of the national income except in time of major war, and two-thirds of that was state and local spending. Federal spending typically amounted to 3 percent or less of the national income. Since 1933 government spending has never been less than 20 percent of national income and is now over 40 percent, and two-thirds of that is spending by the federal government. True, much of the period since the end of World War II has been a period of cold or hot war. However, since 1946 nondefense spending alone has never been less than 16 percent of the national income and is now roughly one-third the national income. Federal government spending alone is more than one-quarter of the national income in total, and more than a fifth for nondefense purposes alone. By this measure the role of the federal government in the economy has multiplied roughly tenfold in the past half-century.

Roosevelt was inaugurated on March 4, 1933—when the economy was at its lowest ebb. Many states had declared a banking holiday, closing their banks. Two days after he was inaugurated, President Roosevelt ordered all banks throughout the nation to close. But Roosevelt used his inaugural address to deliver a message of hope, proclaiming that "the only thing we have to fear is fear itself." And he immediately launched a frenetic program of legislative measures—the "hundred days" of a special congressional session.

The members of FDR's brain trust were drawn mainly from the universities—in particular, Columbia University. They reflected the change that had occurred earlier in the intellectual atmosphere on the campuses—from belief in individual responsibility, laissez-faire, and a decentralized and limited government to belief in social responsibility and a centralized and powerful government. It was the function of government, they believed, to protect individuals from the vicissitudes of fortune and to control the operation of the economy in the "general interest," even if that involved government ownership and operation of the means of production.

These two strands were already present in a famous novel published in 1887, *Looking Backward* by Edward Bellamy, a utopian fantasy in which a Rip Van Winkle character who goes to sleep in the year 1887 awakens in the year 2000 to discover a changed world. "Looking backward," his new companions explain to him how the utopia that astonishes him emerged in the 1930s—a prophetic date—from the hell of the 1880s. That utopia involved the promise of security "from cradle to grave"—the first use of that phrase we have come across—as well as detailed government planning, including compulsory national service by all persons over an extended period.[1]

Coming from this intellectual atmosphere, Roosevelt's advisers were all too ready to view the depression as a failure of capitalism and to believe that active intervention by government—and especially central government—was the appropriate remedy. Benevolent public servants, disinterested experts, should assume the power that narrow-minded, selfish "economic royalists" had abused. In the words of Roosevelt's first inaugural address, "The moneychangers have fled from the high seats in the temple of our civilization."

In designing programs for Roosevelt to adopt, they could draw not only on the campus, but on the earlier experience of Bismarck's Germany, Fabian England, and middle-way Sweden.

The New Deal, as it emerged during the 1930s, clearly reflected these views. It included programs designed to reform the basic structure of the economy. Some of these had to be abandoned when they were declared unconstitutional by the Supreme Court, notably the NRA (National Recovery Administration) and the AAA (Agricultural Adjustment Administration). Others are still with us, notably the Securities and Exchange Commission, the National Labor Relations Board, nationwide minimum wages.

The New Deal also included programs to provide security against misfortune, notably Social Security (OASI: Old Age and Survivors Insurance), unemployment insurance, and public assistance. This chapter discusses these measures and their later progeny.

The New Deal also included programs intended to be strictly temporary, designed to deal with the emergency situation created by the Great Depression. Some of the temporary programs became permanent, as is the way with government programs.

The most important temporary programs included "make work" projects under the Works Progress Administration, the use of unemployed youth to improve the national parks and forests under the Civilian Conservation Corps, and direct federal relief to the indigent. At the time, these programs served a useful function. There was distress on a vast scale; it was important to do something about that distress promptly, both to assist the people in distress and to restore hope and confidence to the public. These programs were hastily contrived, and no doubt were imperfect and wasteful, but that was understandable and unavoidable under the circumstances. The Roosevelt administration achieved a considerable measure of success in relieving immediate distress and restoring confidence.

World War II interrupted the New Deal, while at the same time strengthening greatly its foundations. The war brought massive government budgets and unprecedented control by government over the details of economic life: fixing of prices and wages by edict, rationing of consumer goods, prohibition of the production of some civilian goods, allocation of raw materials and finished products, control of imports and exports.

The elimination of unemployment, the vast production of war matériel that made the United States the "arsenal of democracy," and unconditional victory over Germany and Japan—all these were widely interpreted as demonstrating the capacity of government to run the economic system more effectively than "unplanned capitalism." One of the first pieces of major legislation enacted after the war was the Employment Act of 1946, which expressed government's responsibility for maintaining "maximum employment, production and·purchasing power" and, in effect, enacted Keynesian policies into law.

The war's effect on public attitudes was the mirror image of the depression's. The depression convinced the public that capitalism was defective; the war, that centralized government was efficient. Both conclusions were false. The depression was produced by a failure of government, not of private enterprise. As to the war, it is one thing for government to exercise great control temporarily for a single overriding purpose shared by almost all citizens and for which almost all citizens are willing to make heavy sacrifices; it is a very different thing for government to control the economy permanently to promote a vaguely defined "public interest" shaped by the enormously varied and diverse objectives of its citizens.

At the end of the war it looked as if central economic planning was the wave of the future. That outcome was passionately welcomed by some who saw it as the dawn of a world of plenty shared equally. It was just as passionately feared by others, including us, who saw it as a turn to tyranny and misery. So far, neither the hopes of the one nor the fears of the other have been realized.

Government has expanded greatly. However, that expansion has not taken the form of detailed central economic planning accompanied by ever widening nationalization of industry, finance, and commerce, as so many of us feared it would. Experience put an end to detailed economic planning, partly because it was not successful in achieving the announced objectives, but also because it conflicted with freedom. That conflict was clearly evident in the attempt by the British government to control the jobs people could hold. Adverse public reaction forced the abandonment of the attempt. Nationalized industries proved so inefficient and generated such large losses in Britain, Sweden, France, and the United States that only a few die-hard Marxists today regard further nationalization as desirable. The illusion that nationalization increases productive efficiency, once widely shared, is gone. Additional nationalization does occur—passenger railroad service and some freight service in the United States, Leyland Motors in Great Britain, steel in Sweden. But it occurs for very different reasons—because consumers wish to retain services subsidized by the government when market conditions call for their curtailment or because workers in unprofitable industries fear unemployment. Even the supporters of such nationalization regard it as at best a necessary evil.

The failure of planning and nationalization has not eliminated pressure for an ever bigger government. It has simply altered its direction. The expansion of government now takes the form of welfare programs and of regulatory activities. As W. Allen Wallis put it in a somewhat different context, socialism, "intellectually bankrupt after more than a century of seeing one after another of its arguments for socializing the *means* of production demolished—now seeks to socialize the *results* of production." [2]

In the welfare area the change of direction has led to an explosion in recent decades, especially after President Lyndon Johnson declared a "War on Poverty" in 1964. New Deal programs of Social Security, unemployment insurance, and direct relief were all expanded to cover new groups; payments were increased; and Medicare, Medicaid, food stamps, and numerous other programs were added. Public housing and urban renewal programs were enlarged. By now there are literally hundreds of government welfare and income transfer programs. The Department of Health, Education and Welfare, established in 1953 to consolidate the scattered welfare programs, began with a budget of $2 billion, less than 5 percent of expenditures on national defense. Twenty-five years later, in 1978, its budget was $160 billion, one and a half times as much as total spending on the army, the navy, and the air force. It had the third largest budget in the world, exceeded only by the entire budget of the U.S. government and of the Soviet Union. The department supervised a huge empire, penetrating every corner of the nation. More than one out of every 100 persons employed in this country worked in the HEW empire, either directly for the department or in programs for which HEW had responsibility but which were administered by state or local government units. All of us were affected by its activities. (In late 1979, HEW was subdivided by the creation of a separate Department of Education.)

No one can dispute two superficially contradictory phenomena: widespread dissatisfaction with the results of this explosion in welfare activities; continued pressure for further expansion.

The objectives have all been noble; the results, disappointing. Social Security expenditures have skyrocketed, and the system is in deep financial trouble. Public housing and urban renewal programs have subtracted from rather than added to the housing available to the poor. Public assistance rolls mount despite growing employment. By general agreement, the welfare program is a "mess" saturated with fraud and corruption. As government has paid a larger share of the nation's medical bills, both patients and physicians complain of rocketing costs and of the increasing impersonality of medicine. In education, student performance has dropped as federal intervention has expanded (Chapter 6).

The repeated failure of well-intentioned programs is not an accident. It is not simply the result of mistakes of execution. The failure is deeply rooted in the use of bad means to achieve good objectives.

Despite the failure of these programs, the pressure to expand them grows. Failures are attributed to the miserliness of Congress in appropriating funds, and so are met with a cry for still bigger programs. Special interests that benefit from specific programs press for their expansion—foremost among them the massive bureaucracy spawned by the programs.

An attractive alternative to the present welfare system is a negative income tax. This proposal has been widely supported by individuals and groups of all political persuasions. A variant has been proposed by three Presidents; yet it seems politically unfeasible for the foreseeable future.

THE EMERGENCE OF THE MODERN WELFARE STATE

The first modern state to introduce on a fairly large scale the kind of welfare measures that have become popular in most societies today was the newly created German empire under the leadership of the "Iron Chancellor," Otto von Bismarck. In the early 1880s he introduced a comprehensive scheme of social security, offering the worker insurance against accident, sickness, and old age. His motives were a complex mixture of paternalistic concern for the lower classes and shrewd politics. His measures served to undermine the political appeal of the newly emerging Social Democrats.

It may seem paradoxical that an essentially autocratic and aristocratic state such as pre–World War I Germany—in today's jargon, a right-wing dictatorship—should have led the way in introducing measures that are generally linked to socialism and the Left. But there is no paradox—even putting to one side Bismarck's political motives. Believers in aristocracy and socialism share a faith in centralized rule, in rule by command rather than by voluntary cooperation. They differ in who should rule: whether an elite determined by birth or experts supposedly chosen on merit. Both proclaim, no doubt sincerely, that they wish to promote the well-being of the "general public," that they know what is in the "public interest" and how to attain it better than the ordinary person. Both, therefore, profess a paternalistic philosophy. And both end up, if they attain power, promoting the interests of their own class in the name of the "general welfare."

More immediate precursors of the social security measures adopted in the 1930s were the measures taken in Great Britain beginning with the Old Age Pensions Act passed in 1908 and the National Insurance Act in 1911.

The Old Age Pensions Act granted to any person over the age of seventy whose income fell below a specified sum a weekly pension that varied according to the recipient's income. It was strictly noncontributory, and so was in one sense simply direct relief—an extension of Poor Law provisions that had in one form or another existed in Great Britain for centuries. However, as A. V. Dicey points out, there was a fundamental difference. The pension was regarded as a right whose receipt, in the words of the act, "shall not deprive the pensioner of any franchise, right or privilege, or subject him to any disability." It shows how far we have come from that modest beginning that Dicey, commenting on the act five years after its enactment, could write, "Surely a sensible and a benevolent man may well ask himself whether England as a whole will gain by enacting that the receipt of poor relief, in the shape of a pension, shall be consistent with the pensioner's retaining the right to join in the election of a Member of Parliament." [3] It would take a modern Diogenes with a powerful lamp to find anyone today who could vote if receipt of government largesse were a disqualification.

The National Insurance Act aimed "at the attainment of two objects: The first is that any person . . . who is employed in the United Kingdom . . . shall, from the age of 16 to 70, be insured against ill-health, or in other words, be insured the means for curing illness. . . . The second object is that any such person who is employed in certain employments specified in the Act shall be insured against unemployment, or, in other words, be secured support during periods of unemployment." [4] Unlike old-age pensions, the system established was contributory. It was to be financed partly by employers, partly by employees, partly by the government.

Both because of its contributory nature and because of the contingencies that it sought to insure against, this act was an even more radical departure from prior practice than the Old Age Pensions Act. "[U]nder the National Insurance Act," wrote Dicey,

> the State incurs new and, it may be, very burdensome, duties, and confers upon wage-earners new and very extensive rights. . . . [B]efore 1908 the question whether a man, rich or poor, should insure his health, was a matter left entirely to the free discretion or indiscretion of each individual. His conduct no more concerned the State than the question whether he should wear a black coat or a brown coat.
>
> But the National Insurance Act will, in the long run, bring upon the State, that is, upon the taxpayers, a far heavier responsibility than is anticipated by English electors. . . . [U]nemployment insurance . . . is in fact the admission by a State of its duty to insure a man against the evil ensuing from his having no work. . . . The National Insurance Act is in accordance with the doctrine of socialism, it is hardly reconcilable with the liberalism, or even the radicalism of 1865.[5]

These early British measures, like Bismarck's, illustrate the affinity between aristocracy and socialism. In 1904 Winston Churchill left the Tory party—the party of the aristocracy—for the Liberal party. As a member of Lloyd George's cabinet he took a leading role in social reform legislation. The change of party, which proved temporary, required no change of principles—as it would have a half-century earlier, when the Liberal party was the party of free trade abroad and laissez-faire at home. The social legislation he sponsored, while different in scope and kind, was in the tradition of the paternalistic Factory Acts that had been adopted in the nineteenth century largely under the influence of the so-called Tory Radicals[6]—a group drawn in considerable part from the aristocracy and imbued with a sense of obligation to look after the interests of the working classes, and to do so with their consent and backing, not through coercion.

It is no exaggeration to say that the shape of Britain today owes more to Tory principles of the nineteenth century than to the ideas of Karl Marx and Friedrich Engels.

Another example that doubtless influenced FDR's New Deal was *Sweden, The Middle Way*, as Marquis Childs would title his book, published in 1936. Sweden enacted compulsory old-age pensions in 1915 as a contributory system. Pensions were payable to all after the age of sixty-seven regardless of financial status. The size of the pension depended on the payments individuals had made into the system. Such payments were supplemented by government funds.

In addition to old-age pensions and, later, unemployment insurance, Sweden went in for government ownership of industry, public housing, and consumers' cooperatives on a large scale.

RESULTS OF THE WELFARE STATE

Britain and Sweden, long the two countries most frequently pointed to as successful welfare states, have had increasing difficulties. Dissatisfaction has mounted in both countries.

Britain has found it increasingly difficult to finance growing government spending. Taxes have become a major source of resentment. And resentment has been multiplied manyfold by the impact of inflation (see Chapter 9). The National Health Service, once the prize jewel in the welfare state crown and still widely regarded by much of the British public as one of the great achievements of the Labour government, has run into increasing difficulties—plagued by strikes, rising costs, and lengthening waiting lists of patients. And more and more people have been turning to private physicians, private health insurance, hospitals, and rest homes. Though still a minor sector of the health industry, the private sector has been growing rapidly.

Unemployment in Britain has mounted along with inflation. The government has had to renege on its commitment to full employment. Underlying everything else, productivity and real income in Britain have at best been stagnant, so that Britain has been falling far behind its continental neighbors. The dissatisfaction surfaced dramatically in the Tory party's sizable election victory in 1979, a victory gained on Margaret Thatcher's promise of a drastic change in government direction.

Sweden has done far better than Britain. It was spared the

burden of two world wars and, indeed, reaped economic benefits from its neutrality. Nonetheless, it too has recently been experiencing the same difficulties as Britain: high inflation and high unemployment; opposition to high taxes, resulting in the emigration of some of its most talented people; dissatisfaction with social programs. Here, too, the voters have expressed their views at the ballot box. In 1976 the voters ended over four decades of rule by the Social Democratic party, and replaced it by a coalition of other parties, though as yet there has been no basic change in the direction of government policy.

New York City is the most dramatic example in the United States of the results of trying to do good through government programs. New York is the most welfare-oriented community in the United States. Spending by the city government is larger relative to its population than in any other city in the United States— double that in Chicago. The philosophy that guided the city was expressed by Mayor Robert Wagner in his 1965 budget message: "I do not propose to permit our fiscal problems to set the limits of our commitments to meet the essential needs of the people of the city." [7] Wagner and his successors proceeded to interpret "essential needs" very broadly indeed. But more money, more programs, more taxes didn't work. They led to financial catastrophe without meeting "the essential needs of the people" even on a narrow interpretation, let alone on Wagner's. Bankruptcy was prevented only by assistance from the federal government and the State of New York, in return for which New York City surrendered control over its affairs, becoming a closely supervised ward of state and federal governments.

New Yorkers naturally sought to blame outside forces for their problem, but as Ken Auletta wrote in a recent book, New York "was not compelled to create a vast municipal hospital or City University system, to continue free tuition, institute open enrollment, ignore budget limitations, impose the steepest taxes in the nation, borrow beyond its means, subsidize middle-income housing, continue rigid rent controls, reward municipal workers with lush pension, pay and fringe benefits."

He quips, "Goaded by liberalism's compassion and ideological commitment to the redistribution of wealth, New York officials helped redistribute much of the tax base and thousands of jobs out of New York." [8]

One fortunate circumstance was that New York City has no power to issue money. It could not use inflation as a means of taxation and thus postpone the evil day. Unfortunately, instead of really facing up to its problems, it simply cried for help from the State of New York and the federal government. Let us take a closer look at a few other examples.

Social Security

The major welfare-state program in the United States on the federal level is Social Security—old age, survivors, disability, and health insurance. On the one hand, it is a sacred cow that no politician can question—as Barry Goldwater discovered in 1964. On the other hand, it is the target of complaints from all sides. Persons receiving payments complain that the sums are inadequate to maintain the standard of life they had been led to expect. Persons paying Social Security taxes complain that they are a heavy burden. Employers complain that the wedge introduced by the taxes between the cost to the employer of adding a worker to his payroll and the net gain to the worker of taking a job creates unemployment. Taxpayers complain that the unfunded obligations of the Social Security system total many trillions of dollars, and that not even the present high taxes will keep it solvent for long. And all complaints are justified!

Social Security and unemployment insurance were enacted in the 1930s to enable working people to provide for their own retirement and for temporary periods of unemployment rather than becoming objects of charity. Public assistance was introduced to aid persons in distress, with the expectation that it would be phased out as employment improved and as Social Security took over the task. Both programs started small. Both have grown like Topsy. Social Security has shown no sign of displacing public assistance—both are at all time highs in terms of both dollar expenditures and number of persons receiving payments. In 1978 payments under Social Security for retirement, disability, unemployment, hospital and medical care, and to survivors totaled more than $130 billion and were made to more than 40 million recipients.[9] Public assistance payments of more than $40 billion were made to more than 17 million recipients.

To keep the discussion within manageable limits, we shall restrict this section to the major component of Social Security—old age and survivors' benefits, which accounted for nearly two-thirds of total expenditures and three-quarters of the recipients. The next section deals with public assistance programs.

Social Security was enacted in the 1930s and has been promoted ever since through misleading labeling and deceptive advertising. A private enterprise that engaged in such labeling and advertising would doubtless be severely castigated by the Federal Trade Commission.

Consider the following paragraph that appeared year after year until 1977 in millions of copies of an unsigned HEW booklet entitled *Your Social Security*: "The basic idea of social security is a simple one: During working years employees, their employers, and self-employed people pay social security contributions which are pooled into special trust funds. When earnings stop or are reduced because the worker retires, becomes disabled, or dies, monthly cash benefits are paid to replace part of the earnings the family has lost." [10]

This is Orwellian doublethink.

Payroll taxes are labeled "contributions" (or, as the Party might have put it in the book *Nineteen Eighty-Four*,[11] "Compulsory is Voluntary").

Trust funds are conjured with as if they played an important role. In fact, they have long been extremely small ($32 billion for OASI as of June 1978, or less than half a year's outlays at the current rate) and consist only of promises by one branch of government to pay another branch. The present value of the old-age pensions already promised to persons covered by Social Security (both those who have retired and those who have not) is in the trillions of dollars. That is the size of the trust fund that would be required to justify the words of the booklet (in Orwellian terms, "Little is Much").

The impression is given that a worker's "benefits" are financed by his "contributions." The fact is that taxes collected from persons at work were used to pay benefits to persons who had retired or to their dependents and survivors. No trust fund in any meaningful sense was being accumulated ("I am You").

Workers paying taxes today can derive no assurance from trust funds that they will receive benefits when they retire. Any assurance derives solely from the willingness of future taxpayers to impose taxes on themselves to pay for benefits that present taxpayers are promising themselves. This one-sided "compact between the generations," foisted on generations that cannot give their consent, is a very different thing from a "trust fund." It is more like a chain letter.

The HEW booklets, including those currently being distributed, also say, "Nine out of ten working people in the United States are earning protection for themselves and their families under the social security program." [12]

More doublethink. What nine out of ten working people are now doing is paying taxes to finance payments to persons who are not working. The individual worker is not "earning" protection for himself and his family in the sense in which a person who contributes to a private vested pension system can be said to be "earning" his own protection. He is only "earning" protection in the political sense of satisfying certain administrative requirements for qualifying for benefits. Persons who now receive payments get much more than the actuarial value of the taxes that they paid and that were paid on their behalf. Young persons who now pay Social Security taxes are being promised much less than the actuarial value of the taxes that they will pay and that will be paid on their behalf.

Social Security is in no sense an insurance program in which individual payments purchase equivalent actuarial benefits. As even its strongest supporters admit, "The relationship between individual contributions (that is, payroll taxes) and benefits received is extremely tenuous." [13] Social Security is, rather, a combination of a particular tax and a particular program of transfer payments.

The fascinating thing is that we have never met anyone, whatever his political persuasion, who would defend either the tax system by itself or the benefit system by itself. Had the two components been considered separately, neither would ever have been adopted!

Consider the tax. Except for a recent minor modification (rebates under the earned income credit), it is a flat rate on wages up to a maximum, a tax that is regressive, bearing most heavily on persons with low incomes. It is a tax on work, which discourages employers from hiring workers and discourages people from seeking work.

Consider the benefit arrangement. Payments are determined neither by the amount paid by the beneficiary nor by his financial status. They constitute neither a fair return for prior payments nor an effective way of helping the indigent. There is a link between taxes paid and benefits received, but that is at best a fig leaf to give some semblance of credibility to calling the combination "insurance." The amount of money a person gets depends on all sorts of adventitious circumstances. If he happened to work in a covered industry, he gets a benefit; if he happened to work in a noncovered industry, he does not. If he worked in a covered industry for only a few quarters, he gets nothing, no matter how indigent he may be. A woman who has never worked, but is the wife or widow of a man who qualifies for the maximum benefit, gets precisely the same amount as a similarly situated woman who, in addition, qualifies for benefits on the basis of her own earnings. A person over sixty-five who decides to work and who earns more than a modest amount a year not only gets no benefits but, to add insult to injury, must pay additional taxes—supposedly to finance the benefits that are not being paid. And this list could be extended indefinitely.

We find it hard to conceive of a greater triumph of imaginative packaging than the combination of an unacceptable tax and an unacceptable benefit program into a Social Security program that is widely regarded as one of the greatest achievements of the New Deal.

As we have gone through the literature on Social Security, we have been shocked at the arguments that have been used to defend the program. Individuals who would not lie to their children, their friends, or their colleagues, whom all of us would trust implicitly in the most important personal dealings, have propagated a false view of Social Security. Their intelligence and exposure to contrary views make it hard to believe that they have done so unintentionally and innocently. Apparently they have regarded themselves as an elite group within society that knows what is good for other people better than those people do for themselves, an elite that has a duty and a responsibility to persuade the voters to pass laws that will be good for them, even if they have to fool the voters in order to get them to do so.

The long-run financial problems of Social Security stem from one simple fact: the number of people receiving payments from the system has increased and will continue to increase faster than the number of workers on whose wages taxes can be levied to finance those payments. In 1950 seventeen persons were employed for every person receiving benefits; by 1970 only three; by early in the twenty-first century, if present trends continue, at most two will be.

As these remarks indicate, the Social Security program involves a transfer from the young to the old. To some extent such a transfer has occurred throughout history—the young supporting their parents, or other relatives, in old age. Indeed, in many poor countries with high infant death rates, like India, the desire to assure oneself of progeny who can provide support in old age is a major reason for high birth rates and large families. The difference between Social Security and earlier arrangements is that Social Security is compulsory and impersonal—earlier arrangements were voluntary and personal. Moral responsibility is an individual matter, not a social matter. Children helped their parents out of love or duty. They now contribute to the support of someone else's parents out of compulsion and fear. The earlier transfers strengthened the bonds of the family; the compulsory transfers weaken them.

In addition to the transfer from young to old, Social Security also involves a transfer from the less well-off to the better-off. True, the benefit schedule is biased in favor of persons with lower wages, but this effect is much more than offset by another. Children from poor families tend to start work—and start paying employment taxes—at a relatively early age; children from higher income families at a much later age. At the other end of the life cycle, persons with lower incomes on the average have a shorter life span than persons with higher incomes. The net result is that the poor tend to pay taxes for more years and receive benefits for fewer years than the rich—all in the name of helping the poor! This perverse effect is reinforced by a number of other features of Social Security. The exemption of benefits from income tax is more valuable, the higher the other income of the recipient. The restriction on payments to persons sixty-five to seventy-two (to become seventy in 1982) is based solely on earnings during those years and not on other categories of income—$1 million of dividends does not disqualify anyone from receiving Social Security benefits; wages or salary of more than $4,500 a year produce a loss of $1 of benefits for every $2 of earnings.[14]

All in all, Social Security is an excellent example of Director's Law in operation, namely, "Public expenditures are made for the primary benefit of the middle class, and financed with taxes which are borne in considerable part by the poor and rich." [15]

Public Assistance

We can be far briefer in discussing the "welfare mess" than in discussing Social Security—because on this question there is more agreement. The defects of our present system of welfare have become widely recognized. The relief rolls grow despite growing affluence. A vast bureaucracy is largely devoted to shuffling papers rather than to serving people. Once people get on relief, it is hard to get off. The country is increasingly divided into two classes of citizens, one receiving relief and the other paying for it. Those on relief have little incentive to earn income. Relief payments vary widely from one part of the country to another, which encourages migration from the South and the rural areas to the North, and particularly to urban centers. Persons who are or have been on relief are treated differently from those who have not been on relief (the so-called working poor) though both may be on the same economic level. Public anger is repeatedly stirred by widespread corruption and cheating, well-publicized reports of welfare "queens" driving around in Cadillacs bought with multiple relief checks.

As complaints about welfare programs have mounted, so have the number of programs to be complained about. There is a ragbag of well over 100 federal programs that have been enacted to help the poor. There are major programs like Social Security, unemployment insurance, Medicare, Medicaid, aid to families with dependent children, supplemental security income, food stamps, and myriad minor ones most people have never heard of, such as assistance to Cuban refugees; special supplemental feeding for women, infants, and children; intensive infant care project; rent supplements; urban rat control; comprehensive hemophilia treatment centers; and so on. One program duplicates another. Some families who manage to receive assistance from numerous programs end up with an income decidedly higher than the average income for the country. Other families, through ignorance or apathy, fail to apply for programs that might ease real distress. But every program requires a bureaucracy to administer it.

Over and above the more than $130 billion per year spent under Social Security, expenditure on these programs is around $90 billion a year—ten times the amount spent in 1960. This is clearly overkill. The so-called poverty level for 1978, as estimated by the Census, was close to $7,000 for a nonfarm family of four, and about 25 million persons were said to be members of families below the poverty level. That is a gross overestimate because it classifies families solely by money income, neglecting entirely any income in kind—from an owned home, a garden, food stamps, Medicaid, public housing. Several studies suggest that allowing for these omissions would cut the Census estimates by one-half or three-quarters.[16] But even if you use the Census estimates, they imply that expenditures on welfare programs amounted to about $3,500 per person below the poverty level, or about $14,000 per family of four—roughly twice the poverty level itself. If these funds were all going to the "poor," there would be no poor left—they would be among the comfortably well-off, at least.

Clearly, this money is not going primarily to the poor. Some is siphoned off by administrative expenditures, supporting a massive bureaucracy at attractive pay scales. Some goes to people who by no stretch of the imagination can be regarded as indigent. These are the college students who get food stamps and perhaps other forms of assistance, the families with comfortable incomes who get housing subsidies, and so on in more varied forms than your or our imagination can encompass. Some goes to the welfare cheats.

Yet this much must be said for these programs. Unlike that of Social Security recipients, the average income of the people who are subsidized by these vast sums is probably lower than the average income of the people who pay the taxes to support them —though even that cannot be asserted with certainty. As Martin Anderson put it,

> There may be great inefficiencies in our welfare programs, the level of fraud may be very high, the quality of management may be terrible, the programs may overlap, inequities may abound, and the financial incentive to work may be virtually non-existent. But if we step back and judge the vast array of welfare programs . . . by two basic criteria—the completeness of coverage for those who really need help, and the adequacy of the amount of help they do receive— the picture changes dramatically. Judged by these standards our welfare system has been a brilliant success.[17]

Housing Subsidies

From small beginnings in the New Deal years, government programs to provide housing have expanded rapidly. A new Cabinet department, the Department of Housing and Urban Development, was created in 1965. It now has a staff of nearly 20,000 persons that disburses more than $10 billion a year. Federal housing programs have been supplemented by state and city government programs, especially in New York State and New York City. The programs started with government construction of housing units for low-income families. After the war an urban renewal program was added, and in many areas, public housing was extended to "middle-income" families. More recently "rent supplements"—government subsidization of rents for privately owned housing units—have been added.

In terms of the initial objective, these programs have been a conspicuous failure. More dwelling units were destroyed than were built. Those families who got apartments at subsidized rents benefited. Those families who were forced to move to poorer housing because their homes were destroyed and not replaced were worse off. Housing is better and more widely distributed in the United States today than when the public housing program was started, but that has occurred through private enterprise despite the government subsidies.

The public housing units themselves have frequently become slums and hotbeds of crime, especially juvenile delinquency. The most dramatic case was the Pruitt-Igoe public housing project in St. Louis—a massive apartment complex covering fifty-three acres that won an architectural prize for design. It deteriorated to such an extent that part of it had to be blown up. At that point only 600 of 2,000 units were occupied and the project was said to look like an urban battleground.

We well remember an episode that occurred when we toured the Watts area of Los Angeles in 1968. We were being shown the area by the man who was in charge of a well-run self-help project sponsored by a trade union. When we commented on the attractiveness of some apartment houses in the area, he broke out angrily: "That's the worst thing that ever happened to Watts. That's public housing." He went on to say, "How do you expect youngsters to develop good character and values when they live in a development consisting entirely of broken families, almost all on welfare?" He deplored also the effect of the public housing developments on juvenile delinquency and on the neighborhood schools, which were disproportionately filled with children from broken families.

Recently we heard a similar evaluation of public housing from a leader of a "sweat-equity" housing project in the South Bronx, New York. The area looks like a bombed-out city, with many buildings abandoned as a result of rent control and others destroyed by riots. The "sweat-equity" group has undertaken to rehabilitate an area of these abandoned buildings by their own efforts into housing that they can subsequently occupy. Initially they received outside help only in the form of a few private grants. More recently they have also been receiving some assistance from government.

When we asked our respondent why his group adopted their arduous approach rather than simply moving into public housing, he gave an answer like the one we had heard in Los Angeles, with the added twist that building and owning their own homes would give the participants in the project a sense of pride in their homes that would lead them to maintain them properly.

Part of the government assistance that "sweat-equity" received was the services of CETA workers. These people were paid by the government under the Comprehensive Employment and Training Act and assigned to various public projects to acquire training that it was hoped would enable them to get private jobs. When we asked our respondent whether the sweat-equity group would rather have the help of CETA workers or the money that was being spent on them, he left no doubt whatsoever that they would prefer the money. All in all, it was heartening to observe the sense of self-reliance, independence, and energy displayed on this self-help project by contrast with the apathy, sense of futility, and boredom so evident at public housing projects we visited.

New York's subsidized "middle-income" housing, justified as a way to keep middle-income families from fleeing the city, presents a very different picture. Spacious and luxurious apartments are rented at subsidized rates to families who are "middle-income" only by a most generous use of that term. The apartments are on the average subsidized in the amount of more than $200 per month. "Director's Law" at work again.

Urban renewal was adopted with the aim of eliminating slums —"urban blight." The government subsidized the acquisition and clearance of areas to be renewed and made much of the cleared land available to private developers at artificially low prices. Urban renewal destroyed "four homes, most of them occupied by blacks, for every home it built—most of them to be occupied by middle- and upper-income whites." [18] The original occupants were forced to move elsewhere, often turning another area into a "blighted" one. The program well deserves the names "slum removal" and "Negro removal" that some critics gave it.

The chief beneficiaries of public housing and urban renewal have not been the poor people. The beneficiaries have, rather, been the owners of property purchased for public housing or located in urban renewal areas; middle- and upper-income families who were able to find housing in the high-priced apartments or townhouses that frequently replaced the low-rental housing that was renewed out of existence; the developers and occupants of shopping centers constructed in urban areas; institutions such as universities and churches that were able to use urban renewal projects to improve their neighborhoods. As a recent *Wall Street Journal* editorial put it,

> The Federal Trade Commission has looked into the government's housing policies and discovered that they are driven by something more than pure altruism. An FTC staff policy briefing book finds that the main thrust seems to come from people who make money building housing—contractors, bankers, labor unions, materials suppliers, etc. After the housing is built, the government and these various "constituencies" take less interest in it. So the FTC has been getting complaints about the quality of housing built under federal programs, about leaky roofs, inadequate plumbing, bad foundations, etc.[19]

In the meantime, even where it was not deliberately destroyed, low-priced rental housing deteriorated because of rent control and similar measures.

Medical Care

Medicine is the latest welfare field in which the role of government has been exploding. State and local governments, and to a lesser extent the federal government, have long had a role in public health (sanitation, contagious diseases, etc.) and in provision of hospital facilities. In addition, the federal government has provided medical care for the military and veterans. However, as late as 1960 government expenditures for civilian health purposes (i.e., omitting the military and veterans) were less than $5 billion, or a little over 1 percent of the national income. After the introduction of Medicare and Medicaid in 1965, government spending on health mounted rapidly, reaching $68 billion by 1977, or about 4.5 percent of national income. The government's share of total expenditures on medical care has almost doubled, from 25 percent in 1960 to 42 percent in 1977. The clamor for the federal government to assume a still larger role continues. President Carter has come out in favor of national health insurance, though in a limited form because of financial constraints. Senator Edward M. Kennedy has no such inhibitions; he favors the immediate enactment of complete government responsibility for the health care of the nation's citizens.

Extra government spending has been paralleled by a rapid growth in private health insurance. Total spending on medical care doubled from 1965 to 1977 as a fraction of national income. Medical facilities have expanded, too, but not as rapidly as expenditures. The inevitable result has been sharp increases in the price of medical care and in the incomes of physicians and others engaged in rendering medical services.

The government has responded by trying to regulate the medical procedures followed and to hold down the fees charged by physicians and hospitals. And so it should. If the government spends the taxpayers' money, it is right and proper that it should be concerned with what it gets for what it spends: he who pays the piper calls the tune. If the present trends continue, the end result will inevitably be socialized medicine.

National health insurance is another example of misleading labeling. In such a system there would be no connection between what you would pay and the actuarial value of what you would be entitled to receive, as there is in private insurance. In addition it is not directed at insuring "national health"—a meaningless phrase—but at providing medical services to the residents of the country. What its proponents are in fact proposing is a system of socialized medicine. As Dr. Gunnar Biörck, an eminent Swedish professor of medicine and head of the department of medicine at a major Swedish hospital, has written:

> The setting in which medicine has been practiced during thousands of years has been one in which the *patient* has been the client and employer of the physician. Today the State, in one manifestation or the other, claims to be the employer and, thus, the one to prescribe the conditions under which the physician has to carry out his work. These conditions may not—and will eventually not—be restricted to working hours, salaries and certified drugs; they may invade the whole territory of the patient-physician relationship. . . . If the battle of today is not fought and not won, there will be no battle to fight tomorrow.[20]

Proponents of socialized medicine in the United States—to give their cause its proper name—typically cite Great Britain, and more recently Canada, as examples of its success. The Canadian experience has been too recent to provide an adequate test—most new brooms sweep pretty clean—but difficulties are already emerging. The British National Health Service has now been in operation more than three decades, and the results are pretty conclusive. That, no doubt, is why Canada has been replacing Britain as the example pointed to. A British physician, Dr. Max Gammon, spent five years studying the British Health Service. In a December 1976 report he wrote: "[The National Health Service] brought centralized state financing and control of delivery to virtually all medical services in the country. The voluntary system of financing and delivery of medical care which had been developed in Britain over the preceding 200 years was almost entirely eliminated. The existing compulsory system was reorganized and made practically universal."

Also, "No new hospitals were in fact built in Britain during the first thirteen years of the National Health Service and there are now, in 1976, fewer hospital beds in Britain than in July 1948 when the National Health Service took over." [21]

And, we may add, two-thirds of those beds are in hospitals that were built before 1900 by private medicine and private funds.

Dr. Gammon was led by his survey to promulgate what he calls a theory of bureaucratic displacement: the more bureaucratic an organization, the greater the extent to which useless work tends to displace useful work—an interesting extension of one of Parkinson's laws. He illustrates the theory with hospital services in Britain from 1965 to 1973. In that eight-year period hospital staffs in total increased in number by 28 percent, administrative and clerical help by 51 percent. But output, as measured by the average number of hospital beds occupied daily, actually went *down* by 11 percent. And not, as Dr. Gammon hastened to point out, because of any lack of patients to occupy the beds. At all times there was a waiting list for hospital beds of around 600,000 people. Many must wait for years to have an operation that the health service regards as optional or postponable.

Physicians are fleeing the British Health Service. About one-third as many physicians emigrate each year from Britain to other countries as graduate from its medical schools. The recent rapid growth of strictly private medical practice, private health insurance, and private hospitals and nursing homes is another result of dissatisfaction with the Health Service.

Two major arguments are offered for introducing socialized medicine in the United States: first, that medical costs are beyond the means of most Americans; second, that socialization will somehow reduce costs. The second can be dismissed out of hand —at least until someone can find some example of an activity that is conducted more economically by government than by private enterprise. As to the first, the people of the country must pay the costs one way or another; the only question is whether they pay them directly on their own behalf, or indirectly through the mediation of government bureaucrats who will subtract a substantial slice for their own salaries and expenses.

In any event, the costs of ordinary medical care are well within the means of most American families. Private insurance arrangements are available to meet the contingency of an unusually large expense. Already, 90 percent of all hospital bills are paid through third-party payments. Exceptional hardship cases no doubt arise, and some help, private or public, may well be desirable for them. But help for a few hardship cases hardly justifies putting the whole population in a straitjacket.

To give a sense of proportion, the total expenditures on medical care, private and governmental, amount to less than two-thirds as much as spending on housing, about three-quarters as much as spending on automobiles, and only two and a half times as much as spending on alcohol and tobacco—which undoubtedly adds to medical bills.

In our opinion there is no case whatsoever for socialized medicine. On the contrary, government already plays too large a role in medical care. Any further expansion of its role would be very much against the interests of patients, physicians, and health care personnel. We discuss another aspect of medical care—the licensing of physicians and its bearing on the power of the American Medical Association—in Chapter 8 on "Who Protects the Worker?"

THE FALLACY OF THE WELFARE STATE

Why have all these programs been so disappointing? Their objectives were surely humanitarian and noble. Why have they not been achieved?

At the dawn of the new era all seemed well. The people to be benefited were few; the taxpayers available to finance them, many —so each was paying a small sum that provided significant benefits to a few in need. As welfare programs expanded, the numbers changed. Today all of us are paying out of one pocket to put money—or something money could buy—in the other.

A simple classification of spending shows why that process leads to undesirable results. When you spend, you may spend your own money or someone else's; and you may spend for the benefit of yourself or someone else. Combining these two pairs of alternatives gives four possibilities summarized in the following simple table:[22]

YOU ARE THE SPENDER

On Whom Spent

Whose Money	You	Someone Else
Yours	I	II
Someone Else's	III	IV

Category I in the table refers to your spending your own money on yourself. You shop in a supermarket, for example. You clearly have a strong incentive both to economize and to get as much value as you can for each dollar you do spend.

Category II refers to your spending your own money on someone else. You shop for Christmas or birthday presents. You have the same incentive to economize as in Category I but not the same incentive to get full value for your money, at least as judged by the tastes of the recipient. You will, of course, want to get something the recipient will like—provided that it also makes the right impression and does not take too much time and effort. (If, indeed, your main objective were to enable the recipient to get as much value as possible per dollar, you would give him cash, converting your Category II spending to Category I spending by him.)

Category III refers to your spending someone else's money on yourself—lunching on an expense account, for instance. You have no strong incentive to keep down the cost of the lunch, but you do have a strong incentive to get your money's worth.

Category IV refers to your spending someone else's money on still another person. You are paying for someone else's lunch out of an expense account. You have little incentive either to economize or to try to get your guest the lunch that he will value most highly. However, if you are having lunch with him, so that the lunch is a mixture of Category III and Category IV, you do have a strong incentive to satisfy your own tastes at the sacrifice of his, if necessary.

All welfare programs fall into either Category III—for example, Social Security which involves cash payments that the recipient is free to spend as he may wish; or Category IV—for example, public housing; except that even Category IV programs share one feature of Category III, namely, that the bureaucrats administering the program partake of the lunch; and all Category III programs have bureaucrats among their recipients.

In our opinion these characteristics of welfare spending are the main source of their defects.

Legislators vote to spend someone else's money. The voters who elect the legislators are in one sense voting to spend their own money on themselves, but not in the direct sense of Category I spending. The connection between the taxes any individual pays and the spending he votes for is exceedingly loose. In practice, voters, like legislators, are inclined to regard someone else as paying for the programs the legislator votes for directly and the voter votes for indirectly. Bureaucrats who administer the programs are also spending someone else's money. Little wonder that the amount spent explodes.

The bureaucrats spend someone else's money on someone else. Only human kindness, not the much stronger and more dependable spur of self-interest, assures that they will spend the money in the way most beneficial to the recipients. Hence the wastefulness and ineffectiveness of the spending.

But that is not all. The lure of getting someone else's money is strong. Many, including the bureaucrats administering the programs, will try to get it for themselves rather than have it go to someone else. The temptation to engage in corruption, to cheat, is strong and will not always be resisted or frustrated. People who resist the temptation to cheat will use legitimate means to direct the money to themselves. They will lobby for legislation favorable to themselves, for rules from which they can benefit. The bureaucrats administering the programs will press for better pay and perquisites for themselves—an outcome that larger programs will facilitate.

The attempt by people to divert government expenditures to themselves has two consequences that may not be obvious. First, it explains why so many programs tend to benefit middle- and upper-income groups rather than the poor for whom they are supposedly intended. The poor tend to lack not only the skills valued in the market, but also the skills required to be successful in the political scramble for funds. Indeed, their disadvantage in the political market is likely to be greater than in the economic. Once well-meaning reformers who may have helped to get a welfare measure enacted have gone on to their next reform, the poor are left to fend for themselves and they will almost always be overpowered by the groups that have already demonstrated a greater capacity to take advantage of available opportunities.

The second consequence is that the net gain to the recipients of the transfer will be less than the total amount transferred. If $100 of somebody else's money is up for grabs, it pays to spend up to $100 of your own money to get it. The costs incurred to lobby legislators and regulatory authorities, for contributions to political campaigns, and for myriad other items are a pure waste— harming the taxpayer who pays and benefiting no one. They must be subtracted from the gross transfer to get the net gain—and may, of course, at times exceed the gross transfer, leaving a net loss, not gain.

These consequences of subsidy seeking also help to explain the pressure for more and more spending, more and more programs. The initial measures fail to achieve the objectives of the well-meaning reformers who sponsored them. They conclude that not enough has been done and seek additional programs. They gain as allies both people who envision careers as bureaucrats administering the programs and people who believe that they can tap the money to be spent.

Category IV spending tends also to corrupt the people involved. All such programs put some people in a position to decide what is good for other people. The effect is to instill in the one group a feeling of almost God-like power; in the other, a feeling of childlike dependence. The capacity of the beneficiaries for independence, for making their own decisions, atrophies through disuse. In addition to the waste of money, in addition to the failure to achieve the intended objectives, the end result is to rot the moral fabric that holds a decent society together.

Another by-product of Category III or IV spending has the same effect. Voluntary gifts aside, you can spend someone else's money only by taking it away as government does. The use of force is therefore at the very heart of the welfare state—a bad means that tends to corrupt the good ends. That is also the reason why the welfare state threatens our freedom so seriously.

WHAT SHOULD BE DONE

Most of the present welfare programs should never have been enacted. If they had not been, many of the people now dependent on them would have become self-reliant individuals instead of wards of the state. In the short run that might have appeared cruel for some, leaving them no option to low-paying, unattractive work. But in the long run it would have been far more humane. However, given that the welfare programs exist, they cannot simply be abolished overnight. We need some way to ease the transition from where we are to where we would like to be, of providing assistance to people now dependent on welfare while at the same time encouraging an orderly transfer of people from welfare rolls to payrolls.

Such a transitional program has been proposed that could enhance individual responsibility, end the present division of the nation into two classes, reduce both government spending and the present massive bureaucracy, and at the same time assure a safety net for every person in the country, so that no one need suffer dire distress. Unfortunately, the enactment of such a program seems a utopian dream at present. Too many vested interests—ideological, political, and financial—stand in the way.

Nonetheless, it seems worth outlining the major elements of such a program, not with any expectation that it will be adopted in the near future, but in order to provide a vision of the direction in which we should be moving, a vision that can guide incremental changes.

The program has two essential components: first, reform the present welfare system by replacing the ragbag of specific programs with a single comprehensive program of income supplements in cash—a negative income tax linked to the positive income tax; second, unwind Social Security while meeting present commitments and gradually requiring people to make their own arrangements for their own retirement.

Such a comprehensive reform would do more efficiently and humanely what our present welfare system does so inefficiently and inhumanely. It would provide an assured minimum to all persons in need regardless of the reasons for their need while doing as little harm as possible to their character, their independence, or their incentive to better their own condition.

The basic idea of a negative income tax is simple, once we penetrate the smoke screen that conceals the essential features of the positive income tax. Under the current positive income tax you are permitted to receive a certain amount of income without paying any tax. The exact amount depends on the size of your family, your age, and on whether you itemize your deductions. This amount is composed of a number of elements—personal exemptions, low-income allowance, standard deduction (which has recently been relabeled the zero bracket amount), the sum corresponding to the general tax credit, and for all we know still other items that have been added by the Rube Goldberg geniuses who have been having a field day with the personal income tax. To simplify the discussion, let us use the simpler British term of "personal allowances" to refer to this basic amount.

If your income exceeds your allowances, you pay a tax on the excess at rates that are graduated according to the size of the excess. Suppose your income is less than the allowances? Under the current system, those unused allowances in general are of no value. You simply pay no tax.[23]

If your income happened just to equal your allowances in each of two succeeding years, you would pay no tax in either year. Suppose you had that same income for the two years together, but more than half was received the first year. You would have a positive taxable income, that is, income in excess of allowances for that year, and would pay tax on it. In the second year, you would have a negative taxable income, that is, your allowances would exceed your income but you would, in general, get no benefit from your unused allowances. You would end up paying more tax for the two years together than if the income had been split evenly.[24]

With a negative income tax, you would receive from the government some fraction of the unused allowances. If the fraction you received was the same as the tax rate on the positive income, the total tax you paid in the two years would be the same regardless of how your income was divided between them.

When your income was above allowances, you would pay tax, the amount depending on the tax rates charged on various amounts of income. When your income was below allowances, you would receive a subsidy, the amount depending on the subsidy rates attributed to various amounts of unused allowances.

The negative income tax would allow for fluctuating income, as in our example, but that is not its main purpose. Its main purpose is rather to provide a straightforward means of assuring every family a minimum amount, while at the same time avoiding a massive bureaucracy, preserving a considerable measure of individual responsibility, and retaining an incentive for individuals to work and earn enough to pay taxes instead of receiving a subsidy.

Consider a particular numerical example. In 1978 allowances amounted to $7,200 for a family of four, none above age sixty-five. Suppose a negative income tax had been in existence with a subsidy rate of 50 percent of unused allowances. In that case, a family of four that had no income would have qualified for a subsidy of $3,600. If members of the family had found jobs and earned an income, the amount of the subsidy would have gone down, but the family's total income—subsidy plus earnings—would have gone up. If earnings had been $1,000, the subsidy would have gone down to $3,100 and total income up to $4,100. In effect, the earnings would have been split between reducing the subsidy and raising the family's income. When the family's earnings reached $7,200, the subsidy would have fallen to zero. That would have been the *break-even* point at which the family would have neither received a subsidy nor paid a tax. If earnings had gone still higher, the family would have started paying a tax.

We need not here go into administrative details—whether subsidies would be paid weekly, biweekly, or monthly, how compliance would be checked, and so on. It suffices to say that these questions have all been thoroughly explored; that detailed plans have been developed and submitted to Congress—a matter to which we shall return.

The negative income tax would be a satisfactory reform of our present welfare system only if it *replaces* the host of other specific programs that we now have. It would do more harm than good if it simply became another rag in the ragbag of welfare programs.

If it did replace them, the negative income tax would have enormous advantages. It is directed specifically at the problem of poverty. It gives help in the form most useful to the recipient, namely, cash. It is general—it does not give help because the recipient is old or disabled or sick or lives in a particular area, or any of the other many specific features entitling people to benefits under current programs. It gives help because the recipient has a low income. It makes explicit the cost borne by taxpayers. Like any other measure to alleviate poverty, it reduces the incentive of people who are helped to help themselves. However, if the subsidy rate is kept at a reasonable level, it does not eliminate that incentive entirely. An extra dollar earned always means more money available for spending.

Equally important, the negative income tax would dispense with the vast bureaucracy that now administers the host of welfare programs. A negative income tax would fit directly into our current income tax system and could be administered along with it. It would reduce evasion under the current income tax since everyone would be required to file income tax forms. Some additional personnel might be required, but nothing like the number who are now employed to administer welfare programs.

By dispensing with the vast bureaucracy and integrating the subsidy system with the tax system, the negative income tax would eliminate the present demoralizing situation under which some people—the bureaucrats administering the programs—run other people's lives. It would help to eliminate the present division of the population into two classes—those who pay and those who are supported on public funds. At reasonable break-even levels and tax rates, it would be far less expensive than our present system.

There would still be need for personal assistance to some families who are unable for one reason or another to manage their own affairs. However, if the burden of income maintenance were handled by the negative income tax, that assistance could and would be provided by private charitable activities. We believe that one of the greatest costs of our present welfare system is that it not only undermines and destroys the family, but also poisons the springs of private charitable activity.

Where does Social Security fit into this beautiful, if politically unfeasible, dream?

The best solution in our view would be to combine the enactment of a negative income tax with winding down Social Security while living up to present obligations. The way to do that would be:

1. Repeal immediately the payroll tax.

2. Continue to pay all existing beneficiaries under Social Security the amounts that they are entitled to under current law.

3. Give every worker who has already earned coverage a claim to those retirement, disability, and survivors benefits that his tax payments and earnings to date would entitle him to under current law, reduced by the present value of the reduction in his future taxes as a result of the repeal of the payroll tax. The worker could choose to take his benefits in the form of a future annuity *or* government bonds equal to the present value of the benefits to which he would be entitled.

4. Give every worker who has not yet earned coverage a capital sum (again in the form of bonds) equal to the accumulated

value of the taxes that he or his employer has paid on his behalf.

5. Terminate any further accumulation of benefits, allowing individuals to provide for their own retirement as they wish.

6. Finance payments under items 2, 3, and 4 out of general tax funds plus the issuance of government bonds.

This transition program does not add in any way to the true debt of the U.S. government. On the contrary, it reduces that debt by ending promises to future beneficiaries. It simply brings into the open obligations that are now hidden. It funds what is now unfunded. These steps would enable most of the present Social Security administrative apparatus to be dismantled at once.

The winding down of Social Security would eliminate its present effect of discouraging employment and so would mean a larger national income currently. It would add to personal saving and so lead to a higher rate of capital formation and a more rapid rate of growth of income. It would stimulate the development and expansion of private pension plans and so add to the security of many workers.

WHAT IS POLITICALLY FEASIBLE?

This is a fine dream, but unfortunately it has no chance whatsoever of being enacted at present. Three Presidents—Presidents Nixon, Ford, and Carter—have considered or recommended a program including elements of a negative income tax. In each case political pressures have led them to offer the program as an addition to many existing programs, rather than as a substitute for them. In each case the subsidy rate was so high that the program gave little if any incentive to recipients to earn income. These misshapen programs would have made the whole system worse, not better. Despite our having been the first to have proposed a negative income tax as a replacement for our present welfare system, one of us testified before Congress *against* the version that President Nixon offered as the Family Assistance Plan.[25]

The political obstacles to an acceptable negative income tax are of two related kinds. The more obvious is the existence of vested interests in present programs: the recipients of benefits, state and local officials who regard themselves as benefiting from the programs, and, above all, the welfare bureaucracy that administers them.[26] The less obvious obstacle is the conflict among the objectives that advocates of welfare reform, including existing vested interests, seek to achieve.

As Martin Anderson puts it in an excellent chapter on "The Impossibility of Radical Welfare Reform,"

> All radical welfare reform schemes have three basic parts that are politically sensitive to a high degree. The first is the basic benefit level provided, for example, to a family of four on welfare. The second is the degree to which the program affects the incentive of a person on welfare to find work or to earn more. The third is the additional cost to the taxpayers.
>
> . . . To become a political reality the plan must provide a decent level of support for those on welfare, it must contain strong incentives to work, and it must have a reasonable cost. *And it must do all three at the same time.*[27]

The conflict arises from the content given to "decent," to "strong," and to "reasonable," but especially to "decent." If a "decent" level of support means that few if any current recipients are to receive less from the reformed program than they now do from the collection of programs available, then it is impossible to achieve all three objectives simultaneously, no matter how "strong" and "reasonable" are interpreted. Yet, as Anderson says, "There is no way that the Congress, at least in the near future, is going to pass any kind of welfare reform that actually reduces payments for millions of welfare recipients."

Consider the simple negative income tax that we introduced as an illustration in the preceding section: a break-even point for a family of four of $7,200, a subsidy rate of 50 percent, which means a payment of $3,600 to a family with no other source of support. A subsidy rate of 50 percent would give a tolerably strong incentive to work. The cost would be far less than the cost of the present complex of programs. However, the support level is politically unacceptable today. As Anderson says, "The typical welfare family of four in the United States now [early 1978] qualifies for about $6,000 in services and money every year. In higher paying states, like New York, a number of welfare families receive annual benefits ranging from $7,000 to $12,000 and more." [28]

Even the $6,000 "typical" figure requires a subsidy rate of 83.3 percent if the break-even point is kept at $7,200. Such a rate would both seriously undermine the incentive to work and add enormously to cost. The subsidy rate could be reduced by making the break-even point higher, but that would add greatly to the cost. This is a vicious circle from which there is no escape. So long as it is not politically feasible to reduce the payments to many persons who now receive high benefits from multiple current programs, Anderson is right: "*There is no way to achieve all the politically necessary conditions for radical welfare reform at the same time.*" [29]

However, what is not politically feasible today may become politically feasible tomorrow. Political scientists and economists have had a miserable record in forecasting what will be politically feasible. Their forecasts have repeatedly been contradicted by experience.

Our great and revered teacher Frank H. Knight was fond of illustrating different forms of leadership with ducks that fly in a *V* with a leader in front. Every now and then, he would say, the ducks behind the leader would veer off in a different direction while the leader continued flying ahead. When the leader looked around and saw that no one was following, he would rush to get in front of the *V* again. That is one form of leadership—undoubtedly the most prevalent form in Washington.

While we accept the view that our proposals are not currently feasible politically, we have outlined them as fully as we have, not only as an ideal that can guide incremental reform, but also in the hope that they may, sooner or later, become politically feasible.

CONCLUSION

The empire ruled over until recently by the Department of *Health, Education and Welfare* has been spending more and more of our money each year on our *health*. The main effect has simply been to raise the costs of medical and health services without any corresponding improvement in the quality of medical care.

Spending on *education* has been skyrocketing, yet by common consent the quality of education has been declining. Increasing sums and increasingly rigid controls have been imposed on us to promote racial integration, yet our society seems to be becoming more fragmented.

Billions of dollars are being spent each year on *welfare*, yet at a time when the average standard of life of the American citizen is higher than it has ever been in history, the welfare rolls are growing. The Social Security budget is colossal, yet Social Security is in deep financial trouble. The young complain, and with much justice, about the high taxes they must pay, taxes that are needed to finance the benefits going to the old. Yet the old complain, and with much justice, that they cannot maintain the standard of living that they were led to expect. A program that was enacted to make sure that our older folks never became objects of charity has seen the number of old persons on welfare rolls grow.

By its own accounting, in one year HEW lost through fraud, abuse, and waste an amount of money that would have sufficed to build well over 100,000 houses costing more than $50,000 each.

The waste is distressing, but it is the least of the evils of the paternalistic programs that have grown to such massive size. Their major evil is their effect on the fabric of our society. They weaken the family; reduce the incentive to work, save, and innovate; reduce the accumulation of capital; and limit our freedom. These are the fundamental standards by which they should be judged.

CHAPTER 5

Created
Equal

"Equality," "liberty"—what precisely do these words from the Declaration of Independence mean? Can the ideals they express be realized in practice? Are equality and liberty consistent one with the other, or are they in conflict?

Since well before the Declaration of Independence, these questions have played a central role in the history of the United States. The attempt to answer them has shaped the intellectual climate of opinion, led to bloody war, and produced major changes in economic and political institutions. This attempt continues to dominate our political debate. It will shape our future as it has our past.

In the early decades of the Republic, equality meant equality before God; liberty meant the liberty to shape one's own life. The obvious conflict between the Declaration of Independence and the institution of slavery occupied the center of the stage. That conflict was finally resolved by the Civil War. The debate then moved to a different level. Equality came more and more to be interpreted as "equality of opportunity" in the sense that no one should be prevented by arbitrary obstacles from using his capacities to pursue his own objectives. That is still its dominant meaning to most citizens of the United States.

Neither equality before God nor equality of opportunity presented any conflict with liberty to shape one's own life. Quite the opposite. Equality and liberty were two faces of the same basic value—that every individual should be regarded as an end in himself.

A very different meaning of equality has emerged in the United States in recent decades—equality of outcome. Everyone should have the same level of living or of income, should finish the race at the same time. Equality of outcome is in clear conflict with liberty. The attempt to promote it has been a major source of bigger and bigger government, and of government-imposed restrictions on our liberty.

EQUALITY BEFORE GOD

When Thomas Jefferson, at the age of thirty-three, wrote "all men are created equal," he and his contemporaries did not take these words literally. They did not regard "men"—or as we would say today, "persons"—as equal in physical characteristics, emotional reactions, mechanical and intellectual abilities. Thomas Jefferson himself was a most remarkable person. At the age of twenty-six he designed his beautiful house at Monticello (Italian for "little mountain"), supervised its construction, and, indeed, is said to have done some of the work himself. In the course of his life, he was an inventor, a scholar, an author, a statesman, governor of the State of Virginia, President of the United States, Minister to France, founder of the University of Virginia—hardly an average man.

The clue to what Thomas Jefferson and his contemporaries meant by equal is in the next phrase of the Declaration—"endowed by their Creator with certain unalienable rights; that among these are Life, Liberty, and the pursuit of Happiness." Men were equal before God. Each person is precious in and of himself. He has unalienable rights, rights that no one else is entitled to invade. He is entitled to serve his own purposes and not to be treated simply as an instrument to promote someone else's purposes. "Liberty" is part of the definition of equality, not in conflict with it.

Equality before God—personal equality[1]—is important precisely because people are not identical. Their different values, their different tastes, their different capacities will lead them to want to lead very different lives. Personal equality requires respect for their right to do so, not the imposition on them of someone else's values or judgment. Jefferson had no doubt that some men were superior to others, that there was an elite. But that did not give them the right to rule others.

If an elite did not have the right to impose its will on others, neither did any other group, even a majority. Every person was to be his own ruler—provided that he did not interfere with the similar right of others. Government was established to protect that right—from fellow citizens and from external threat—not to give a majority unbridled rule. Jefferson had three achievements he wanted to be remembered for inscribed on his tombstone: the Virginia statute for religious freedom (a precursor of the U.S. Bill of Rights designed to protect minorities against domination by majorities), authorship of the Declaration of Independence, and the founding of the University of Virginia. The goal of the framers of the Constitution of the United States, drafted by Jefferson's contemporaries, was a national government strong enough to defend the country and promote the general welfare but at the same time sufficiently limited in power to protect the individual citizen, and the separate state governments, from domination by the national government. Democratic, in the sense of widespread participation in government, yes; in the political sense of majority rule, clearly no.

Similarly, Alexis de Tocqueville, the famous French political philosopher and sociologist, in his classic *Democracy in America*, written after a lengthy visit in the 1830s, saw equality, not majority rule, as the outstanding characteristic of America. "In America," he wrote,

> the aristocratic element has always been feeble from its birth; and if at the present day it is not actually destroyed, it is at any rate so completely disabled, that we can scarcely assign to it any degree of influence on the course of affairs. The democratic principle, on the contrary, has gained so much strength by time, by events, and by legislation, as to have become not only predominant but all-powerful. There is no family or corporate authority. . . .
>
> America, then, exhibits in her social state a most extraordinary phenomenon. Men are there seen on a greater equality in point of fortune and intellect, or, in other words, more equal in their strength, than in any other country of the world, or in any age of which history has preserved the remembrance.[2]

Tocqueville admired much of what he observed, but he was by no means an uncritical admirer, fearing that democracy carried too far might undermine civic virtue. As he put it, "There is . . . a manly and lawful passion for equality which incites men to wish all to be powerful and honored. This passion tends to elevate the humble to the rank of the great; but there exists also in the human heart a depraved taste for equality, which impels the weak to attempt to lower the powerful to their own level, and reduces men to prefer equality in slavery to inequality with freedom." [3]

It is striking testimony to the changing meaning of words that in recent decades the Democratic party of the United States has been the chief instrument for strengthening that government power which Jefferson and many of his contemporaries viewed as the greatest threat to democracy. And it has striven to increase government power in the name of a concept of "equality" that is almost the opposite of the concept of equality Jefferson identified with liberty and Tocqueville with democracy.

Of course the practice of the founding fathers did not always correspond to their preaching. The most obvious conflict was slavery. Thomas Jefferson himself owned slaves until the day he died—July 4, 1826. He agonized repeatedly about slavery, suggested in his notes and correspondence plans for eliminating slavery, but never publicly proposed any such plans or campaigned against the institution.

Yet the Declaration he drafted had either to be blatantly violated by the nation he did so much to create and form, or slavery had to be abolished. Little wonder that the early decades of the Republic saw a rising tide of controversy about the institution of slavery. That controversy ended in a civil war that, in the words of Abraham Lincoln's Gettysburg Address, tested whether a "nation, conceived in liberty and dedicated to the proposition that all men are created equal . . . can long endure." The nation endured, but only at a tremendous cost in lives, property, and social cohesion.

EQUALITY OF OPPORTUNITY

Once the Civil War abolished slavery and the concept of personal equality—equality before God and the law—came closer to realization, emphasis shifted, in intellectual discussion and in government and private policy, to a different concept—equality of opportunity.

Literal equality of opportunity—in the sense of "identity"—is impossible. One child is born blind, another with sight. One child has parents deeply concerned about his welfare who provide a background of culture and understanding; another has dissolute, improvident parents. One child is born in the United States, another in India, or China, or Russia. They clearly do not have identical opportunities open to them at birth, and there is no way that their opportunities can be made identical.

Like personal equality, equality of opportunity is not to be interpreted literally. Its real meaning is perhaps best expressed by the French expression dating from the French Revolution: *Une carrière ouverte aux les talents*—a career open to the talents. No arbitrary obstacles should prevent people from achieving those positions for which their talents fit them and which their values lead them to seek. Not birth, nationality, color, religion, sex, nor any other irrelevant characteristic should determine the opportunities that are open to a person—only his abilities.

On this interpretation, equality of opportunity simply spells out in more detail the meaning of personal equality, of equality before the law. And like personal equality, it has meaning and importance precisely because people are different in their genetic and cultural characteristics, and hence both want to and can pursue different careers.

Equality of opportunity, like personal equality, is not inconsistent with liberty; on the contrary, it is an essential component of liberty. If some people are denied access to particular positions in life for which they are qualified simply because of their ethnic background, color, or religion, that is an interference with their right to "Life, Liberty, and the pursuit of Happiness." It denies equality of opportunity and, by the same token, sacrifices the freedom of some for the advantage of others.

Like every ideal, equality of opportunity is incapable of being fully realized. The most serious departure was undoubtedly with respect to the blacks, particularly in the South but in the North as well. Yet there was also tremendous progress for blacks and for other groups. The very concept of a "melting pot" reflected the goal of equality of opportunity. So also did the expansion of "free" education at elementary, secondary, and higher levels—though, as we shall see in the next chapter, this development has not been an unmixed blessing.

The priority given to equality of opportunity in the hierarchy of values generally accepted by the public after the Civil War is manifested particularly in economic policy. The catchwords were free enterprise, competition, laissez-faire. Everyone was to be free to go into any business, follow any occupation, buy any property, subject only to the agreement of the other parties to the transaction. Each was to have the opportunity to reap the benefits if he succeeded, to suffer the costs if he failed. There were to be no arbitrary obstacles. Performance, not birth, religion, or nationality, was the touchstone.

One corollary was the development of what many who regarded themselves as the cultural elite sneered at as vulgar materialism—an emphasis on the almighty dollar, on wealth as both the symbol and the seal of success. As Tocqueville pointed out, this emphasis reflected the unwillingness of the community to accept the traditional criteria in feudal and aristocratic societies, namely birth and parentage. Performance was the obvious alternative, and the accumulation of wealth was the most readily available measure of performance.

Another corollary, of course, was an enormous release of human energy that made America an increasingly productive and dynamic society in which social mobility was an everyday reality. Still another, perhaps surprisingly, was an explosion in charitable activity. This explosion was made possible by the rapid growth in wealth. It took the form it did—of nonprofit hospitals, privately endowed colleges and universities, a plethora of charitable organizations directed to helping the poor—because of the dominant values of the society, including, especially, promotion of equality of opportunity.

Of course, in the economic sphere as elsewhere, practice did not always conform to the ideal. Government *was* kept to a minor role; no major obstacles to enterprise were erected, and by the end of the nineteenth century, positive government measures, especially the Sherman Anti-Trust Law, were adopted to eliminate private barriers to competition. But extralegal arrangements continued to interfere with the freedom of individuals to enter various businesses or professions, and social practices unquestionably gave special advantages to persons born in the "right" families, of the "right" color, and practicing the "right" religion. However, the rapid rise in the economic and social position of various less privileged groups demonstrates that these obstacles were by no means insurmountable.

In respect of government measures, one major deviation from free markets was in foreign trade, where Alexander Hamilton's *Report on Manufactures* had enshrined tariff protection for domestic industries as part of the American way. Tariff protection was inconsistent with thoroughgoing equality of opportunity (see Chapter 2) and, indeed, with the free immigration of persons, which was the rule until World War I, except only for Orientals. Yet it could be rationalized both by the needs of national defense and on the very different ground that equality stops at the water's edge—an illogical rationalization that is adopted also by most of today's proponents of a very different concept of equality.

EQUALITY OF OUTCOME

That different concept, equality of outcome, has been gaining ground in this century. It first affected government policy in Great Britain and on the European continent. Over the past half-century it has increasingly affected government policy in the United States as well. In some intellectual circles the desirability of equality of outcome has become an article of religious faith: everyone should finish the race at the same time. As the Dodo said in *Alice in Wonderland*, "*Everybody* has won, and *all* must have prizes."

For this concept, as for the other two, "equal" is not to be interpreted literally as "identical." No one really maintains that everyone, regardless of age or sex or other physical qualities, should have identical rations of each separate item of food, clothing, and so on. The goal is rather "fairness," a much vaguer notion—indeed, one that it is difficult, if not impossible, to define precisely. "Fair shares for all" is the modern slogan that has replaced Karl Marx's, "To each according to his needs, from each according to his ability."

This concept of equality differs radically from the other two. Government measures that promote personal equality or equality of opportunity enhance liberty; government measures to achieve "fair shares for all" reduce liberty. If what people get is to be determined by "fairness," who is to decide what is "fair"? As a chorus of voices asked the Dodo, "But who is to give the prizes?" "Fairness" is not an objectively determined concept once it departs from identity. "Fairness," like "needs," is in the eye of the beholder. If all are to have "fair shares," someone or some group of people must decide what shares are fair—and they must be able to impose their decisions on others, taking from those who have more than their "fair" share and giving to those who have less. Are those who make and impose such decisions equal to those for whom they decide? Are we not in George Orwell's *Animal Farm*, where "all animals are equal, but some animals are more equal than others"?

In addition, if what people get is determined by "fairness" and not by what they produce, where are the "prizes" to come from? What incentive is there to work and produce? How is it to be decided who is to be the doctor, who the lawyer, who the garbage collector, who the street sweeper? What assures that people will accept the roles assigned to them and perform those roles in accordance with their abilities? Clearly, only force or the threat of force will do.

The key point is not merely that practice will depart from the ideal. Of course it will, as it does with respect to the other two concepts of equality as well. The point is rather that there is a fundamental conflict between the *ideal* of "fair shares" or of its precursor, "to each according to his needs," and the *ideal* of personal liberty. This conflict has plagued every attempt to make equality of outcome the overriding principle of social organization. The end result has invariably been a state of terror: Russia, China, and, more recently, Cambodia offer clear and convincing evidence. And even terror has not equalized outcomes. In every case, wide inequality persists by any criterion; inequality between the rulers and the ruled, not only in power, but also in material standards of life.[4]

The far less extreme measures taken in Western countries in the name of equality of outcome have shared the same fate to a lesser extent. They, too, have restricted individual liberty. They, too, have failed to achieve their objective. It has proved impossible to define "fair shares" in a way that is generally acceptable, or to satisfy the members of the community that they are being treated "fairly." On the contrary, dissatisfaction has mounted with every additional attempt to implement equality of outcome.

Much of the moral fervor behind the drive for equality of outcome comes from the widespread belief that it is not fair that some children should have a great advantage over others simply because they happen to have wealthy parents. Of course it is not fair. However, unfairness can take many forms. It can take the form of the inheritance of property—bonds and stocks, houses, factories; it can also take the form of the inheritance of talent—musical ability, strength, mathematical genius. The inheritance of property can be interfered with more readily than the inheritance of talent. But from an ethical point of view, is there any difference between the two? Yet many people resent the inheritance of property but not the inheritance of talent.

Look at the same issue from the point of view of the parent. If you want to assure your child a higher income in life, you can do so in various ways. You can buy him (or her) an education that will equip him to pursue an occupation yielding a high income; or you can set him up in a business that will yield a higher income than he could earn as a salaried employee; or you can leave him property, the income from which will enable him to live better. Is there any ethical difference among these three ways of using your property? Or again, if the state leaves you any money to spend over and above taxes, should the state permit you to spend it on riotous living but not to leave it to your children?

The ethical issues involved are subtle and complex. They are not to be resolved by such simplistic formulas as "fair shares for all." Indeed, if we took that seriously, youngsters with less musical skill should be given the greatest amount of musical training in order to compensate for their inherited disadvantage, and those with greater musical aptitude should be prevented from having access to good musical training; and similarly with all other categories of inherited personal qualities. That might be "fair" to the youngsters lacking in talent, but would it be "fair" to the talented, let alone to those who had to work to pay for training the youngsters lacking talent, or to the persons deprived of the benefits that might have come from the cultivation of the talents of the gifted? Life is not fair. It is tempting to believe that government can rectify what nature has spawned. But it is also important to recognize how much we benefit from the very unfairness we deplore.

There's nothing fair about Marlene Dietrich's having been born with beautiful legs that we all want to look at; or about Muhammad Ali's having been born with the skill that made him a great fighter. But on the other side, millions of people who have enjoyed looking at Marlene Dietrich's legs or watching one of Muhammad Ali's fights have benefited from nature's unfairness in producing a Marlene Dietrich and a Muhammad Ali. What kind of a world would it be if everyone were a duplicate of everyone else?

It is certainly not fair that Muhammad Ali should be able to earn millions of dollars in one night. But wouldn't it have been even more unfair to the people who enjoyed watching him if, in the pursuit of some abstract ideal of equality, Muhammad Ali had not been permitted to earn more for one night's fight—or for each day spent in preparing for a fight—than the lowest man on the totem pole could get for a day's unskilled work on the docks? It might have been possible to do that, but the result would have been to deny people the opportunity to watch Muhammad Ali. We doubt very much that he would have been willing to undergo the arduous regimen of training that preceded his fights, or to subject himself to the kind of fights he has had, if he were limited to the pay of an unskilled dockworker.

Still another facet of this complex issue of fairness can be illustrated by considering a game of chance, for example, an evening at baccarat. The people who choose to play may start the evening with equal piles of chips, but as the play progresses, those piles will become unequal. By the end of the evening, some will be big winners, others big losers. In the name of the ideal of equality, should the winners be required to repay the losers? That would take all the fun out of the game. Not even the losers would like that. They might like it for the one evening, but would they come back again to play if they knew that whatever happened, they'd end up exactly where they started?

This example has a great deal more to do with the real world than one might at first suppose. Every day each of us makes decisions that involve taking a chance. Occasionally it's a big chance —as when we decide what occupation to pursue, whom to marry, whether to buy a house or make a major investment. More often it's a small chance, as when we decide what movie to go to, whether to cross the street against the traffic, whether to buy one security rather than another. Each time the question is, who is to decide what chances we take? That in turn depends on who bears the consequences of the decision. If we bear the consequences, we can make the decision. But if someone else bears the consequences, should we or will we be permitted to make the decision? If you play baccarat as an agent for someone else with his money, will he, or should he, permit you unlimited scope for decision making? Is he not almost certain to set some limit to your discretion? Will he not lay down some rules for you to observe? To take a very different example, if the government (i.e., your fellow taxpayers) assumes the costs of flood damage to your house, can you be permitted to decide freely whether to build your house on a floodplain? It is no accident that increasing government intervention into personal decisions has gone hand in hand with the drive for "fair shares for all."

The system under which people make their own choices—and bear most of the consequences of their decisions—is the system that has prevailed for most of our history. It is the system that gave the Henry Fords, the Thomas Alva Edisons, the George Eastmans, the John D. Rockefellers, the James Cash Penneys the incentive to transform our society over the past two centuries. It is the system that gave other people an incentive to furnish venture capital to finance the risky enterprises that these ambitious inventors and captains of industry undertook. Of course, there were many losers along the way—probably more losers than winners. We don't remember their names. But for the most part they went in with their eyes open. They knew they were taking chances. And win or lose, society as a whole benefited from their willingness to take a chance.

The fortunes that this system produced came overwhelmingly from developing new products or services, or new ways of producing products or services, or of distributing them widely. The resulting addition to the wealth of the community as a whole, to the well-being of the masses of the people, amounted to many times the wealth accumulated by the innovators. Henry Ford acquired a great fortune. The country acquired a cheap and reliable means of transportation and the techniques of mass production. Moreover, in many cases the private fortunes were largely devoted in the end to the benefit of society. The Rockefeller, Ford, and Carnegie foundations are only the most prominent of the numerous private benefactions which are so outstanding a consequence of the operation of a system that corresponded to "equality of opportunity" and "liberty" as these terms were understood until recently.

One limited sample may give the flavor of the outpouring of philanthropic activity in the nineteenth and early twentieth century. In a book devoted to "cultural philanthropy in Chicago from the 1880's to 1917," Helen Horowitz writes:

> At the turn of the century, Chicago was a city of contradictory impulses: it was both a commercial center dealing in the basic commodities of an industrial society and a community caught in the winds of cultural uplift. As one commentator put it, the city was "a strange combination of pork and Plato."
>
> A major manifestation of Chicago's drive toward culture was the establishment of the city's great cultural institutions in the 1880's and early 1890's (the Art Institute, the Newberry Library, the Chicago Symphony Orchestra, the University of Chicago, the Field Museum, the Crerar Library). . . .
>
> These institutions were a new phenomenon in the city. Whatever the initial impetus behind their founding, they were largely organized, sustained, and controlled by a group of businessmen. . . . Yet while privately supported and managed, the institutions were designed for the whole city. Their trustees had turned to cultural philanthropy not so much to satisfy personal aesthetic or scholarly yearnings as to accomplish social goals. Disturbed by social forces they could not control and filled with idealistic notions of culture, these businessmen saw in the museum, the library, the symphony orchestra, and the university a way to purify their city and to generate a civic renaissance.[5]

Philanthropy was by no means restricted to cultural institutions. There was, as Horowitz writes in another connection, "a kind of explosion of activity on many different levels." And Chicago was not an isolated case. Rather, as Horowitz puts it, "Chicago seemed

to epitomize America." [6] The same period saw the establishment
of Hull House in Chicago under Jane Addams, the first of many
settlement houses established throughout the nation to spread
culture and education among the poor and to assist them in their
daily problems. Many hospitals, orphanages, and other charitable
agencies were set up in the same period.

There is no inconsistency between a free market system and
the pursuit of broad social and cultural goals, or between a free
market system and compassion for the less fortunate, whether
that compassion takes the form, as it did in the nineteenth century, of private charitable activity, or, as it has done increasingly
in the twentieth, of assistance through government—provided that
in both cases it is an expression of a desire to help others. There
is all the difference in the world, however, between two kinds of
assistance through government that seem superficially similar:
first, 90 percent of us agreeing to impose taxes on ourselves in
order to help the bottom 10 percent, and second, 80 percent
voting to impose taxes on the top 10 percent to help the bottom
10 percent—William Graham Sumner's famous example of B
and C deciding what D shall do for A.[7] The first may be wise
or unwise, an effective or an ineffective way to help the disadvantaged—but it is consistent with belief in both equality of opportunity and liberty. The second seeks equality of outcome and is
entirely antithetical to liberty.

WHO FAVORS EQUALITY OF OUTCOME?

There is little support for the goal of equality of outcome despite
the extent to which it has become almost an article of religious
faith among intellectuals and despite its prominence in the
speeches of politicians and the preambles of legislation. The talk
is belied alike by the behavior of government, of the intellectuals
who most ardently espouse egalitarian sentiments, and of the public at large.

For government, one obvious example is the policy toward lotteries and gambling. New York State—and particularly New York
City—is widely and correctly regarded as a stronghold of egalitarian sentiment. Yet the New York State government conducts

lotteries and provides facilities for off-track betting on races. It advertises extensively to induce its citizens to buy lottery tickets and bet on the races—at terms that yield a very large profit to the government. At the same time it tries to suppress the "numbers" game, which, as it happens, offers better odds than the government lottery (especially when account is taken of the greater ease of avoiding tax on winnings). Great Britain, a stronghold, if not the birthplace, of egalitarian sentiment, permits private gambling clubs and betting on races and other sporting events. Indeed, wagering is a national pastime and a major source of government income.

For intellectuals, the clearest evidence is their failure to practice what so many of them preach. Equality of outcome can be promoted on a do-it-yourself basis. First, decide exactly what you mean by equality. Do you want to achieve equality within the United States? In a selected group of countries as a whole? In the world as a whole? Is equality to be judged in terms of income per person? Per family? Per year? Per decade? Per lifetime? Income in the form of money alone? Or including such nonmonetary items as the rental value of an owned home; food grown for one's own use; services rendered by members of the family not employed for money, notably the housewife? How are physical and mental handicaps or advantages to be allowed for?

However you decide these issues, you can, if you are an egalitarian, estimate what money income would correspond to your concept of equality. If your actual income is higher than that, you can keep that amount and distribute the rest to people who are below that level. If your criterion were to encompass the world—as most egalitarian rhetoric suggests it should—something less than, say, $200 a year (in 1979 dollars) per person would be an amount that would correspond to the conception of equality that seems implicit in most egalitarian rhetoric. That is about the average income per person worldwide.

What Irving Kristol has called the "new class"—government bureaucrats, academics whose research is supported by government funds or who are employed in government financed "think-tanks," staffs of the many so-called "general interest" or "public policy" groups, journalists and others in the communications industry—are among the most ardent preachers of the doctrine of equality. Yet they remind us very much of the old, if unfair, saw about the Quakers: "They came to the New World to do good, and ended up doing well." The members of the new class are in general among the highest paid persons in the community. And for many among them, preaching equality and promoting or administering the resulting legislation has proved an effective means of achieving such high incomes. All of us find it easy to identify our own welfare with the welfare of the community.

Of course, an egalitarian may protest that he is but a drop in the ocean, that he would be willing to redistribute the excess of his income over his concept of an equal income if everyone else were compelled to do the same. On one level this contention that compulsion would change matters is wrong—even if everyone else did the same, his specific contribution to the income of others would still be a drop in the ocean. His individual contribution would be just as large if he were the only contributor as if he were one of many. Indeed, it would be more valuable because he could target his contribution to go to the very worst off among those he regards as appropriate recipients. On another level compulsion would change matters drastically: the kind of society that would emerge if such acts of redistribution were voluntary is altogether different—and, by our standards, infinitely preferable —to the kind that would emerge if redistribution were compulsory.

Persons who believe that a society of enforced equality is preferable can also practice what they preach. They can join one of the many communes in this country and elsewhere, or establish new ones. And, of course, it is entirely consistent with a belief in personal equality or equality of opportunity and liberty that any group of individuals who wish to live in that way should be free to do so. Our thesis that support for equality of outcome is word-deep receives strong support from the small number of persons who have wished to join such communes and from the fragility of the communes that have been established.

Egalitarians in the United States may object that the fewness of communes and their fragility reflect the opprobrium that a predominantly "capitalist" society visits on such communes and the resulting discrimination to which they are subjected. That may be true for the United States but as Robert Nozick[8] has pointed out, there is one country where that is not true, where, on the contrary, egalitarian communes are highly regarded and prized. That country is Israel. The kibbutz played a major role in early Jewish settlement in Palestine and continues to play an important role in the state of Israel. A disproportionate fraction of the leaders of the Israeli state were drawn from the kibbutzim. Far from being a source of disapproval, membership in a kibbutz confers social status and commands approbation. Everyone is free to join or leave a kibbutz, and kibbutzim have been viable social organizations. Yet at no time, and certainly not today, have more than about 5 percent of the Jewish population of Israel chosen to be members of a kibbutz. That percentage can be regarded as an upper estimate of the fraction of people who would voluntarily choose a system enforcing equality of outcome in preference to a system characterized by inequality, diversity, and opportunity.

Public attitudes about graduated income taxes are more mixed. Recent referenda on the introduction of graduated state income taxes in some states that do not have them, and on an increase in the extent of graduation in other states, have generally been defeated. On the other hand, the federal income tax is highly graduated, at least on paper, though it also contains a large number of provisions ("loopholes") that greatly reduce the extent of graduation in practice. On this showing, there is at least public tolerance of a moderate amount of redistributive taxation.

However, we venture to suggest that the popularity of Reno, Las Vegas, and now Atlantic City is no less faithful an indication of the preferences of the public than the federal income tax, the editorials in the *New York Times* and the *Washington Post*, and the pages of the *New York Review of Books*.

CONSEQUENCES OF EGALITARIAN POLICIES

In shaping our own policy, we can learn from the experience of Western countries with which we share a common intellectual and cultural background, and from which we derive many of our values. Perhaps the most instructive example is Great Britain, which led the way in the nineteenth century toward implementing equality of opportunity and in the twentieth toward implementing equality of outcome.

Since the end of World War II, British domestic policy has been dominated by the search for greater equality of outcome. Measure after measure has been adopted designed to take from the rich and give to the poor. Taxes were raised on income until they reached a top rate of 98 percent on property income and 83 percent on "earned" income, and were supplemented by ever heavier taxes on inheritances. State-provided medical, housing, and other welfare services were greatly expanded, along with payments to the unemployed and the aged. Unfortunately, the results have been very different from those that were intended by the people who were quite properly offended by the class structure that dominated Britain for centuries. There has been a vast redistribution of wealth, but the end result is not an equitable distribution.

Instead, new classes of privileged have been created to replace or supplement the old: the bureaucrats, secure in their jobs, protected against inflation both when they work and when they retire; the trade unions that profess to represent the most downtrodden workers but in fact consist of the highest paid laborers in the land—the aristocrats of the labor movement; and the new millionaires—people who have been cleverest at finding ways around the laws, the rules, the regulations that have poured from Parliament and the bureaucracy, who have found ways to avoid paying taxes on their income and to get their wealth overseas beyond the grasp of the tax collectors. A vast reshuffling of income and wealth, yes; greater equity, hardly.

The drive for equality in Britain failed, not because the wrong measures were adopted—though some no doubt were; not because they were badly administered—though some no doubt were; not because the wrong people administered them—though no doubt some did. The drive for equality failed for a much more fundamental reason. It went against one of the most basic instincts of all human beings. In the words of Adam Smith, "The uniform, constant, and uninterrupted effort of every man to better his condition" [9]—and, one may add, the condition of his children and his children's children. Smith, of course, meant by "condition" not merely material well-being, though certainly that was one component. He had a much broader concept in mind, one that included all of the values by which men judge their success—in particular the kind of social values that gave rise to the outpouring of philanthropic activities in the nineteenth century.

When the law interferes with people's pursuit of their own values, they will try to find a way around. They will evade the law, they will break the law, or they will leave the country. Few of us believe in a moral code that justifies forcing people to give up much of what they produce to finance payments to persons they do not know for purposes they may not approve of. When the law contradicts what most people regard as moral and proper, they will break the law—whether the law is enacted in the name of a noble ideal such as equality or in the naked interest of one group at the expense of another. Only fear of punishment, not a sense of justice and morality, will lead people to obey the law.

When people start to break one set of laws, the lack of respect for the law inevitably spreads to all laws, even those that everyone regards as moral and proper—laws against violence, theft, and vandalism. Hard as it may be to believe, the growth of crude criminality in Britain in recent decades may well be one consequence of the drive for equality.

In addition, that drive for equality has driven out of Britain some of its ablest, best-trained, most vigorous citizens, much to the benefit of the United States and other countries that have given them a greater opportunity to use their talents for their own benefit. Finally, who can doubt the effect that the drive for equality has had on efficiency and productivity? Surely, that is one of the main reasons why economic growth in Britain has fallen so far behind its continental neighbors, the United States, Japan, and other nations over the past few decades.

We in the United States have not gone as far as Britain in promoting the goal of equality of outcome. Yet many of the same consequences are already evident—from a failure of egalitarian measures to achieve their objectives, to a reshuffling of wealth that by no standards can be regarded as equitable, to a rise in criminality, to a depressing effect on productivity and efficiency.

CAPITALISM AND EQUALITY

Everywhere in the world there are gross inequities of income and wealth. They offend most of us. Few can fail to be moved by the contrast between the luxury enjoyed by some and the grinding poverty suffered by others.

In the past century a myth has grown up that free market capitalism—equality of opportunity as we have interpreted that term—increases such inequalities, that it is a system under which the rich exploit the poor.

Nothing could be further from the truth. Wherever the free market has been permitted to operate, wherever anything approaching equality of opportunity has existed, the ordinary man has been able to attain levels of living never dreamed of before. Nowhere is the gap between rich and poor wider, nowhere are the rich richer and the poor poorer, than in those societies that do not permit the free market to operate. That is true of feudal societies like medieval Europe, India before independence, and much of modern South America, where inherited status determines position. It is equally true of centrally planned societies, like Russia or China or India since independence, where access to government determines position. It is true even where central planning was introduced, as in all three of these countries, in the name of equality.

Russia is a country of two nations: a small privileged upper class of bureaucrats, Communist party officials, technicians; and a great mass of people living little better than their great-grandparents did. The upper class has access to special shops, schools, and luxuries of all kind; the masses are condemned to enjoy little more than the basic necessities. We remember asking a tourist guide in Moscow the cost of a large automobile that we saw and being told, "Oh, those aren't for sale; they're only for the Politburo." Several recent books by American journalists document in great detail the contrast between the privileged life of the upper classes and the poverty of the masses.[10] Even on a simpler level, it is noteworthy that the average wage of a foreman is a larger multiple of the average wage of an ordinary worker in a Russian factory than in a factory in the United States —and no doubt he deserves it. After all, an American foreman only has to worry about being fired; a Russian foreman also has to worry about being shot.

China, too, is a nation with wide differences in income—between the politically powerful and the rest; between city and countryside; between some workers in the cities and other workers. A perceptive student of China writes that "the inequality between rich and poor regions in China was more acute in 1957 than in any of the larger nations of the world except perhaps Brazil." He quotes another scholar as saying, "These examples suggest that the Chinese industrial wage structure is not significantly more egalitarian than that of other countries." And he concludes his examination of equality in China, "How evenly distributed would China's income be today? Certainly, it would not be as even as Taiwan's or South Korea's. . . . On the other hand, income distribution in China is obviously more even than in Brazil or South America. . . . We must conclude that China is far from being a society of complete equality. In fact, income differences in China may be quite a bit greater than in a number of countries commonly associated with 'fascist' elites and exploited masses." [11]

Industrial progress, mechanical improvement, all of the great wonders of the modern era have meant relatively little to the wealthy. The rich in Ancient Greece would have benefited hardly at all from modern plumbing: running servants replaced running water. Television and radio—the patricians of Rome could enjoy the leading musicians and actors in their home, could have the leading artists as domestic retainers. Ready-to-wear clothing, supermarkets—all these and many other modern developments would have added little to their life. They would have welcomed the improvements in transportation and in medicine, but for the rest, the great achievements of Western capitalism have redounded primarily to the benefit of the ordinary person. These achievements have made available to the masses conveniences and amenities that were previously the exclusive prerogative of the rich and powerful.

In 1848 John Stuart Mill wrote: "Hitherto it is questionable if all the mechanical inventions yet made have lightened the day's toil of any human being. They have enabled a greater population to live the same life of drudgery and imprisonment, and an increased number of manufacturers and others to make fortunes. They have increased the comforts of the middle classes. But they have not yet begun to effect those great changes in human destiny, which it is in their nature and in their futurity to accomplish." [12]

No one could say that today. You can travel from one end of the industrialized world to the other and almost the only people you will find engaging in backbreaking toil are people who are doing it for sport. To find people whose day's toil has not been lightened by mechanical invention, you must go to the noncapitalist world: to Russia, China, India or Bangladesh, parts of Yugoslavia; or to the more backward capitalist countries—in Africa, the Mideast, South America; and until recently, Spain or Italy.

CONCLUSION

A society that puts equality—in the sense of equality of outcome—ahead of freedom will end up with neither equality nor freedom. The use of force to achieve equality will destroy freedom, and the force, introduced for good purposes, will end up in the hands of people who use it to promote their own interests.

On the other hand, a society that puts freedom first will, as a happy by-product, end up with both greater freedom and greater equality. Though a by-product of freedom, greater equality is not an accident. A free society releases the energies and abilities of people to pursue their own objectives. It prevents some people from arbitrarily suppressing others. It does not prevent some people from achieving positions of privilege, but so long as freedom is maintained, it prevents those positions of privilege from becoming institutionalized; they are subject to continued attack by other able, ambitious people. Freedom means diversity but also mobility. It preserves the opportunity for today's disadvantaged to become tomorrow's privileged and, in the process, enables almost everyone, from top to bottom, to enjoy a fuller and richer life.

CHAPTER 6

What's Wrong
with Our Schools?

Education has always been a major component of the American Dream. In Puritan New England, schools were quickly established, first as an adjunct of the church, later taken over by secular authorities. After the opening of the Erie Canal, the farmers who left the rocky hills of New England for the fertile plains of the Middle West established schools wherever they went, not only primary and secondary schools, but also seminaries and colleges. Many of the immigrants who streamed over the Atlantic in the second half of the nineteenth century had a thirst for education. They eagerly seized the opportunities available to them in the metropolises and large cities where they mostly settled.

At first, schools were private and attendance strictly voluntary. Increasingly, government came to play a larger role, first by contributing to financial support, later by establishing and administering government schools. The first compulsory attendance law was enacted by Massachusetts in 1852, but attendance did not become compulsory in all states until 1918. Government control was primarily local until well into the twentieth century. The neighborhood school, and control by the local school board, was the rule. Then a so-called reform movement got under way, particularly in the big cities, sparked by the wide differences in the ethnic and social composition of different school districts and by the belief that professional educators should play a larger role. That movement gained additional ground in the 1930s along with the general tendency toward both expansion and centralization of government.

We have always been proud, and with good reason, of the widespread availability of schooling to all and the role that public schooling has played in fostering the assimilation of newcomers into our society, preventing fragmentation and divisiveness, and enabling people from different cultural and religious backgrounds to live together in harmony.

Unfortunately, in recent years our educational record has become tarnished. Parents complain about the declining quality of the schooling their children receive. Many are even more disturbed about the dangers to their children's physical well-being. Teachers complain that the atmosphere in which they are required to teach is often not conducive to learning. Increasing numbers of teachers are fearful about their physical safety, even in the classroom. Taxpayers complain about growing costs. Hardly anyone maintains that our schools are giving the children the tools they need to meet the problems of life. Instead of fostering assimilation and harmony, our schools are increasingly a source of the very fragmentation that they earlier did so much to prevent.

At the elementary and secondary level, the quality of schooling varies tremendously: outstanding in some wealthy suburbs of major metropolises, excellent or reasonably satisfactory in many small towns and rural areas, incredibly **bad** in the inner cities of major metropolises.

"The education, or rather the *un*education, of black children from low income families is undoubtedly the greatest disaster area in public education and its most devastating failure. This is doubly tragic for it has always been the official ethic of public schooling that it was the poor and the oppressed who were its greatest beneficiaries." [1]

Public education is, we fear, suffering from the same malady as are so many of the programs discussed in the preceding and subsequent chapters. More than four decades ago Walter Lippmann diagnosed it as "the sickness of an over-governed society," the change from "the older faith . . . that the exercise of unlimited power by men with limited minds and self-regarding prejudices is soon oppressive, reactionary, and corrupt, . . . that the very condition of progress was the limitation of power to the capacity and the virtue of rulers" to the newer faith "that there are no limits to man's capacity to govern others and that, therefore, no limitations ought to be imposed upon government." [2]

For schooling, this sickness has taken the form of denying many parents control over the kind of schooling their children receive either directly, through choosing and paying for the schools their children attend, or indirectly, through local political activity. Power has instead gravitated to professional educators. The sickness has been aggravated by increasing centralization and bureaucratization of schools, especially in the big cities.

Private market arrangements have played a greater role at the college and university level than at the elementary and secondary level. But this sector has not been immune from the sickness of an overgoverned society. In 1928 fewer students were enrolled in government institutions of higher education than in private institutions; by 1978 close to four times as many were. Direct government financing grew less rapidly than government operation because of tuition charges paid by students, but even so, by 1978 direct government grants accounted for more than half of the total expenditures on higher education by all institutions, government and private.

The increased role of government has had many of the same adverse effects on higher education as on elementary and secondary education. It has fostered an atmosphere that both dedicated teachers and serious students often find inimical to learning.

ELEMENTARY AND SECONDARY EDUCATION: THE PROBLEM

Even in the earliest years of the Republic, not only the cities but almost every town and village and most rural districts had schools. In many states or localities, the maintenance of a "common school" was mandated by law. But the schools were mostly privately financed by fees paid by the parents. Some supplementary finance was generally also available from the local, county, or state government, both to pay fees for children whose parents were regarded as unable to do so and to supplement fees paid by parents. Though schooling was neither compulsory nor free, it was practically universal (slaves, of course, excepted). In his report for 1836, the superintendent of common schools of the State of New York asserted: "Under any view of the subject it is reasonable to believe, that in the common schools, private schools and academies, the number of children actually receiving instruction is equal to the whole number between five and sixteen years of age." [3] Conditions doubtless varied from state to state, but by all accounts schooling was widely available to (white) children from families at all economic levels.

Beginning in the 1840s, a campaign developed to replace the diverse and largely private system by a system of so-called free schools, i.e., schools in which parents and others paid the cost indirectly by taxes rather than directly by fees. According to E. G. West, who has studied extensively the development of government's role in schooling, this campaign was not led by dissatisfied parents, but "mainly by teachers and government officials." [4] The most famous crusader for free schools was Horace Mann, "the father of American public education," as he is termed in the *Encyclopaedia Britannica* article on his life.[5] Mann was the first secretary of the Massachusetts State Board of Education established in 1837, and for the next twelve years he conducted an energetic campaign for a school system paid for by government and controlled by professional educators. His main arguments were that education was so important that government had a duty to provide education to every child, that schools should be secular and include children of all religious, social, and ethnic backgrounds, and that universal, free schooling would enable children to overcome the handicaps of the poverty of their parents. "In his secretarial reports to the Massachusetts Board of Education, Mann proclaimed repetitively . . . that education was a good public investment and increased output." [6] Though the arguments were all pitched in terms of the public interest, much of the support of teachers and administrators for the public school movement derived from a narrow self-interest. They expected to enjoy greater certainty of employment, greater assurance that their salaries would be paid, and a greater degree of control if government rather than parents were the immediate paymaster.

"Despite vast difficulties and vigorous opposition . . . the main outlines of" the kind of system urged by Mann "were achieved by the middle of the 19th century." [7] Ever since, most children have attended government schools. A few have continued to attend so-called private schools, mostly schools operated by the Catholic Church and other religious denominations.

The United States was not unique in moving from a mostly

private to a mostly governmental system of schools. Indeed, one authority has described "the gradual acceptance of the view that education ought to be a responsibility of the state" as the "most significant" of the general trends of the nineteenth century "that were still influencing education in all western countries in the second half of the 20th century." [8] Interestingly enough, this trend began in Prussia in 1808, and in France, under Napoleon, about the same time. Britain was even later than the United States. "[U]nder the spell of laissez faire [it] hesitated a long time before allowing the state to intervene in educational affairs," but finally, in 1870, a system of government schools was established, though elementary education was not made compulsory until 1880, and fees were not generally abolished until 1891.[9] In Britain, as in the United States, schooling was almost universal before the government took it over. Professor West has maintained persuasively that the government takeover in Britain, as in the United States, resulted from pressure by teachers, administrators, and well-meaning intellectuals, rather than parents. He concludes that the government takeover reduced the quality and diversity of schooling.[10]

Education is still another example, like Social Security, of the common element in authoritarian and socialist philosophies. Aristocratic and authoritarian Prussia and Imperial France were the pioneers in state control of education. Socialistically inclined intellectuals in the United States, Britain, and later Republican France were the major supporters of state control in their countries.

The establishment of the school system in the United States as an island of socialism in a free market sea reflected only to a very minor extent the early emergence among intellectuals of a distrust of the market and of voluntary exchange. Mostly, it simply reflected the importance that was attached by the community to the ideal of equality of opportunity. The ability of Horace Mann and his associates to tap that deep sentiment enabled them to succeed in their crusade.

Needless to say, the public school system was not viewed as "socialist" but simply as "American." The most important factor determining how the system operated was its decentralized political structure. The U.S. Constitution narrowly limited the powers of the federal government, so that it played no significant role. The states mostly left control of schools to the local community, the town, the small city, or a subdivision of a large city. Close monitoring of the political authorities running the school system by parents was a partial substitute for competition and assured that any widely shared desires of parents were implemented.

Before the Great Depression the situation was already changing. School districts were consolidated, educational districts enlarged, and more and more power was granted to professional educators. After the depression, when the public joined the intellectuals in an unbridled faith in the virtues of government, and especially of central government, the decline of the one-room school and the local school board became a rout. Power shifted rapidly from the local community to broader entities—the city, the county, the state, and more recently, the federal government.

In 1920 local funds made up 83 percent of all revenues of public schools, federal grants less than 1 percent. By 1940 the local share had fallen to 68 percent. Currently it is less than one-half. The state provided most of the rest of the money: 16 percent in 1920, 30 percent in 1940, and currently more than 40 percent. The federal government's share is still small but growing rapidly: from less than 2 percent in 1940 to roughly 8 percent currently.

As professional educators have taken over, control by parents has weakened. In addition, the function assigned to schools has changed. They are still expected to teach the three R's and to transmit common values. In addition, however, schools are now regarded as means of promoting social mobility, racial integration, and other objectives only distantly related to their fundamental task.

In Chapter 4 we referred to the Theory of Bureaucratic Displacement that Dr. Max Gammon had developed after studying the British National Health Service: in his words, in "a bureaucratic system . . . *increase in expenditure* will be matched by *fall in production*. . . . Such systems will act rather like 'black holes' in the economic universe, simultaneously sucking in resources, and shrinking in terms of 'emitted' production." [11]

His theory applies in full force to the effect of the increasing bureaucratization and centralization of the public school system in the United States. In the five years from school year 1971–72 to school year 1976–77, total professional staff in all U.S. public schools went up 8 percent, cost per pupil went up 58 percent in dollars (11 percent after correction for inflation). *Input clearly up.*

The number of students went *down* 4 percent, the number of schools went *down* 4 percent. And we suspect that few readers will demur from the proposition that the quality of schooling went *down* even more drastically than the quantity. That is certainly the story told by the declining grades recorded on standardized examinations. *Output clearly down.*

Is the decline in output per unit of input due to increasingly bureaucratic and centralized organization? As some evidence, the number of school districts went down by 17 percent in the seven-year period from 1970–71 to 1977–78—continuing the longer-term trend to greater centralization. As to bureaucratization, for a somewhat earlier five-year period for which data are available (1968–69 to 1973–74), when the number of students went up 1 percent, the total professional staff went up 15 percent, and teachers 14 percent, *but supervisors went up 44 percent.*[12]

The problem in schooling is not mere size, not simply that school districts have become larger, and that, on the average, each school has more students. After all, in industry, size has often proved a source of greater efficiency, lower cost, and improved quality. Industrial development in the United States gained a great deal from the introduction of mass production, from what economists call the "economies of scale." Why should schooling be different?

It isn't. The difference is not between schooling and other activities but between arrangements under which the consumer is free to choose and arrangements under which the producer is in the saddle so the consumer has little to say. If the consumer is free to choose, an enterprise can grow in size only if it produces an item that the consumer prefers because of either its quality or its price. And size alone will not enable any enterprise to impose a product on the consumer that the consumer does not consider is worth its price. The large size of General Motors has not prevented it from flourishing. The large size of W. T. Grant & Co.

did not save it from bankruptcy. When the consumer is free to choose, size will survive only if it is efficient.

In political arrangements size generally docs affect consumers' freedom to choose. In small communities the individual citizen feels that he has, and indeed does have, more control over what the political authorities do than in large communities. He may not have the same freedom to choose that he has in deciding whether to buy something or not, but at least he has a considerable opportunity to affect what happens. In addition, when there are many small communities, the individual can choose where to live. Of course, that is a complex choice, involving many elements. Nonetheless, it does mean that local governments must provide their citizens with services they regard as worth the taxes they pay or either be replaced or suffer a loss of taxpayers.

The situation is very different when power is in the hands of a central government. The individual citizen feels that he has, and indeed does have, little control over the distant and impersonal political authorities. The possibility of moving to another community, though it may still be present, is far more limited.

In schooling, the parent and child are the consumers, the teacher and school administrator the producers. Centralization in schooling has meant larger size units, a reduction in the ability of consumers to choose, and an increase in the power of producers. Teachers, administrators, and union officials are no different from the rest of us. They may be parents, too, sincerely desiring a fine school system. However, their interests as teachers, as administrators, as union officials are different from their interests as parents and from the interests of the parents whose children they teach. Their interests may be served by greater centralization and bureaucratization even if the interests of the parents are not—indeed, one way in which those interests are served is precisely by reducing the power of parents.

The same phenomenon is present whenever government bureaucracy takes over at the expense of consumer choice: whether in the post office, in garbage collection, or in the many examples in other chapters.

In schooling, those of us who are in the upper-income classes retain our freedom to choose. We can send our children to private schools, in effect paying twice for their schooling—once in taxes to support the public school system, once in school fees. Or we can choose where to live on the basis of the quality of the public school system. Excellent public schools tend to be concentrated in the wealthier suburbs of the larger cities, where parental control remains very real.[13]

The situation is worst in the inner cities of the larger metropolises—New York, Chicago, Los Angeles, Boston. The people who live in these areas can pay twice for their children's schooling only with great difficulty—though a surprising number do so by sending their children to parochial schools. They cannot afford to move to the areas with good public schools. Their only recourse is to try to influence the political authorities who are in charge of the public schools, usually a difficult if not hopeless task, and one for which they are not well qualified. The residents of the inner cities are probably more disadvantaged in respect of the level of schooling they can get for their children than in any other area of life with the possible exception of crime protection—another "service" that is provided by government.

The tragedy, and irony, is that a system dedicated to enabling all children to acquire a common language and the values of U.S. citizenship, to giving all children equal educational opportunity, should in practice exacerbate the stratification of society and provide highly unequal educational opportunity. Expenditures on schooling per pupil are often as high in the inner cities as in even the wealthy suburbs, but the quality of schooling is vastly lower. In the suburbs almost all of the money goes for education; in the inner cities much of it must go to preserving discipline, preventing vandalism, or repairing its effects. The atmosphere in some inner city schools is more like that of a prison than of a place of learning. The parents in the suburbs are getting far more value for their tax dollars than the parents in the inner cities.

A VOUCHER PLAN FOR ELEMENTARY
AND SECONDARY SCHOOLING

Schooling, even in the inner cities, does not have to be the way it is. It was not that way when parents had greater control. It is not that way now where parents still have control.

The strong American tradition of voluntary action has provided many excellent examples that demonstrate what can be done when parents have greater choice. One example at the elementary level is a parochial school, St. John Chrysostom's, that we visited in one of the poorest neighborhoods in New York City's Bronx. Its funds come in part from a voluntary charitable organization, New York's Inner City Scholarship Fund, in part from the Catholic Church, in part from fees. The youngsters at the school are there because their parents chose it. Almost all are from poor families, yet their parents are all paying at least some of the costs. The children are well behaved, eager to learn. The teachers are dedicated. The atmosphere is quiet and serene.

The cost per pupil is far less than in public schools even after account is taken of the free services of those teachers who are nuns. Yet on the average, the children are two grades ahead of their peers in public school. That's because teachers and parents are free to choose how the children shall be taught. Private money has replaced tax money. Control has been taken away from bureaucrats and put back where it belongs.

Another example, this one at the secondary level, is in Harlem. In the 1960s Harlem was devastated by riots. Many teenagers dropped out of school. Groups of concerned parents and teachers decided to do something about it. They used private funds to take over empty stores and they set up what became known as storefront schools. One of the first and most successful was called Harlem Prep, designed to appeal to youngsters for whom conventional education had failed.

Harlem Prep had inadequate physical facilities. Many of its teachers did not have the right pieces of paper to qualify for certification to teach in public schools. But that did not keep them from doing a good job. Though many students had been misfits and dropouts, they found the sort of teaching they wanted at Harlem Prep.

The school was phenomenally successful. Many of its students went to college, including some of the leading colleges. But unfortunately, this story has an unhappy ending. After the initial period of crisis had passed, the school ran short of cash. The Board of Education offered Ed Carpenter (the head of the school and one of its founders) the money, provided he would conform to their regulations. After a long battle to preserve independence, he gave in. The school was taken over by bureaucrats. "I felt," commented Mr. Carpenter, "that a school like Harlem Prep would certainly die, and not prosper, under the rigid bureaucracy of a Board of Education. . . . We had to see what was going to happen. I didn't believe it was going to be good. I am right. What has happened since we have come to the Board of Education is not all good. It is not all bad, but it's more bad than good."

Private ventures of this kind are valuable. However, at best they only scratch the surface of what needs to be done.

One way to achieve a major improvement, to bring learning back into the classroom, especially for the currently most disadvantaged, is to give all parents greater control over their children's schooling, similar to that which those of us in the upper-income classes now have. Parents generally have both greater interest in their children's schooling and more intimate knowledge of their capacities and needs than anyone else. Social reformers, and educational reformers in particular, often self-righteously take for granted that parents, especially those who are poor and have little education themselves, have little interest in their children's education and no competence to choose for them. That is a gratuitous insult. Such parents have frequently had limited opportunity to choose. However, U.S. history has amply demonstrated that, given the opportunity, they have often been willing to sacrifice a great deal, and have done so wisely, for their children's welfare.

No doubt, some parents lack interest in their children's schooling or the capacity and desire to choose wisely. However, they are in a small minority. In any event, our present system unfortunately does little to help their children.

One simple and effective way to assure parents greater freedom to choose, while at the same time retaining present sources of finance, is a voucher plan. Suppose your child attends a public elementary or secondary school. On the average, countrywide, it cost the taxpayer—you and me—about $2,000 per year in 1978 for every child enrolled. If you withdraw your child from a public school and send him to a private school, you save taxpayers about $2,000 per year—but you get no part of that saving except as it is passed on to all taxpayers, in which case it would amount to at most a few cents off your tax bill. You have to pay private tuition in addition to taxes—a strong incentive to keep your child in a public school.

Suppose, however, the government said to you: "If you relieve us of the expense of schooling your child, you will be given a voucher, a piece of paper redeemable for a designated sum of money, if, and only if, it is used to pay the cost of schooling your child at an approved school." The sum of money might be $2,000, or it might be a lesser sum, say $1,500 or $1,000, in order to divide the saving between you and the other taxpayers. But whether the full amount or the lesser amount, it would remove at least a part of the financial penalty that now limits the freedom of parents to choose.[14]

The voucher plan embodies exactly the same principle as the GI bills that provide for educational benefits to military veterans. The veteran gets a voucher good only for educational expense and he is completely free to choose the school at which he uses it, provided that it satisfies certain standards.

Parents could, and should, be permitted to use the vouchers not only at private schools but also at other public schools—and not only at schools in their own district, city, or state, but at any school that is willing to accept their child. That would both give every parent a greater opportunity to choose and at the same time require public schools to finance themselves by charging tuition (wholly, if the voucher corresponded to the full cost; at least partly, if it did not). The public schools would then have to compete both with one another and with private schools.

This plan would relieve no one of the burden of taxation to pay for schooling. It would simply give parents a wider choice as to the form in which their children get the schooling that the community has obligated itself to provide. The plan would also not affect the present standards imposed on private schools in order for attendance at them to satisfy the compulsory attendance laws.

We regard the voucher plan as a partial solution because it affects neither the financing of schooling nor the compulsory attendance laws. We favor going much farther. Offhand, it would appear that the wealthier a society and the more evenly distributed is income within it, the less reason there is for government to finance schooling. The parents bear most of the cost in any event, and the cost for equal quality is undoubtedly higher when they bear the cost indirectly through taxes than when they pay for schooling directly—unless schooling is very different from other government activities. Yet in practice, government financing has accounted for a larger and larger share of total educational expenses as average income in the United States has risen and income has become more evenly distributed.

We conjecture that one reason is the government operation of schools, so that the desire of parents to spend more on schooling as their incomes rose found the path of least resistance to be an increase in the amount spent on government schools. One advantage of a voucher plan is that it would encourage a gradual move toward greater direct parental financing. The desire of parents to spend more on schooling could readily take the form of adding to the amount provided by the voucher. Public financing for hardship cases might remain, but that is a far different matter than having the government finance a school system for 90 percent of the children going to school because 5 or 10 percent of them might be hardship cases.

The compulsory attendance laws are the justification for government control over the standards of private schools. But it is far from clear that there is any justification for the compulsory attendance laws themselves. Our own views on this have changed over time. When we first wrote extensively a quarter of a century ago on this subject, we accepted the need for such laws on the ground that "a stable democratic society is impossible without a minimum degree of literacy and knowledge on the part of most citizens." [15] We continue to believe that, but research that has been done in the interim on the history of schooling in the United States, the United Kingdom, and other countries has persuaded us that compulsory attendance at schools is not necessary to achieve that minimum standard of literacy and knowledge. As already noted, such research has shown that schooling was well-nigh universal in the United States before attendance was required. In the United Kingdom, schooling was well-nigh universal before either compulsory attendance or government financing of schooling existed. Like most laws, compulsory attendance laws have costs as well as benefits. We no longer believe the benefits justify the costs.

We realize that these views on financing and attendance laws will appear to most readers to be extreme. That is why we only state them here to keep the record straight without seeking to support them at length. Instead, we return to the voucher plan— a much more moderate departure from present practice.

Currently, the only widely available alternative to a local public school is a parochial school. Only churches have been in a position to subsidize schooling on a large scale and only subsidized schooling can compete with "free" schooling. (Try selling a product that someone else is giving away!) The voucher plan would produce a much wider range of alternatives—unless it was sabotaged by excessively rigid standards for "approval." The choice among public schools themselves would be greatly increased. The size of a public school would be determined by the number of customers it attracted, not by politically defined geographical boundaries or by pupil assignment. Parents who organized nonprofit schools, as a few families have, would be assured of funds to pay the costs. Voluntary organizations—ranging from vegetarians to Boy Scouts to the YMCA—could set up schools and try to attract customers. And most important, new sorts of private schools could arise to tap the vast new market.

Let us consider briefly some possible problems with the voucher plan and some objections that have been raised to it.

(1) *The church-state issue.* If parents could use their vouchers to pay tuition at parochial schools, would that violate the First Amendment? Whether it does or not, is it desirable to adopt a policy that might strengthen the role of religious institutions in schooling?

The Supreme Court has generally ruled against state laws providing assistance to parents who send their children to parochial schools, although it has never had occasion to rule on a full-fledged voucher plan covering both public and nonpublic schools. However it might rule on such a plan, it seems clear that the Court would accept a plan that excluded church-connected schools but applied to all other private and public schools. Such a restricted plan would be far superior to the present system, and might not be much inferior to a wholly unrestricted plan. Schools now connected with churches could qualify by subdividing themselves into two parts: a secular part reorganized as an independent school eligible for vouchers, and a religious part reorganized as an after-school or Sunday activity paid for directly by parents or church funds.

The constitutional issue will have to be settled by the courts. But it is worth emphasizing that vouchers would go to *parents, not to schools*. Under the GI bills, veterans have been free to attend Catholic or other colleges and, so far as we know, no First Amendment issue has ever been raised. Recipients of Social Security and welfare payments are free to buy food at church bazaars and even to contribute to the collection plate from their government subsidies, with no First Amendment question being asked.

Indeed, we believe that the penalty that is now imposed on parents who do not send their children to public schools violates the spirit of the First Amendment, whatever lawyers and judges may decide about the letter. Public schools teach religion, too—not a formal, theistic religion, but a set of values and beliefs that constitute a religion in all but name. The present arrangements abridge the religious freedom of parents who do not accept the religion taught by the public schools yet are forced to pay to have their children indoctrinated with it, and to pay still more to have their children escape indoctrination.

(2) *Financial cost.* A second objection to the voucher plan is that it would raise the total cost to taxpayers of schooling—because of the cost of vouchers given for the roughly 10 percent of children who now attend parochial and other private schools. That is a "problem" only to those who disregard the present discrimination against parents who send their children to nonpublic schools. Universal vouchers would end the inequity of using tax funds to school some children but not others.

In any event, there is a simple and straightforward solution: let the amount of the voucher be enough less than the current cost per public school child to keep total public expenditures the same. The smaller amount spent in a private competitive school would very likely provide a higher quality of schooling than the larger amount now spent in government schools. Witness the drastically lower cost per child in parochial schools. (The fact that elite, luxury schools charge high tuition is no counter argument, any more than the $12.25 charged by the "21" Club for its Hamburger Twenty-One in 1979 meant that McDonald's could not sell a hamburger profitably for 45 cents and a Big Mac for $1.05.)

(3) *The possibility of fraud.* How can one make sure that the voucher is spent for schooling, not diverted to beer for papa and clothes for mama? The answer is that the voucher would have to be spent in an *approved* school or teaching establishment and could be redeemed for cash only by such schools. That would not prevent *all* fraud—perhaps in the forms of "kickbacks" to parents —but it should keep fraud to a tolerable level.

(4) *The racial issue.* Voucher plans were adopted for a time in a number of southern states to avoid integration. They were ruled unconstitutional. Discrimination under a voucher plan can be prevented at least as easily as in public schools by redeeming vouchers only from schools that do not discriminate. A more difficult problem has troubled some students of vouchers. That is the possibility that voluntary choice with vouchers might increase racial and class separation in schools and thus exacerbate racial conflict and foster an increasingly segregated and hierarchical society.

We believe that the voucher plan would have precisely the opposite effect; it would moderate racial conflict and promote a society in which blacks and whites cooperate in joint objectives, while respecting each other's separate rights and interests. Much objection to forced integration reflects not racism but more or less well-founded fears about the physical safety of children and the quality of their schooling. Integration has been most successful when it has resulted from choice, not coercion. Nonpublic schools, parochial and other, have often been in the forefront of the move toward integration.

Violence of the kind that has been rising in public schools is possible only because the victims are compelled to attend the schools that they do. Give them effective freedom to choose and students—black and white, poor and rich, North and South— would desert schools that could not maintain order. Discipline is seldom a problem in private schools that train students as radio and television technicians, typists and secretaries, or for myriad other specialties.

Let schools specialize, as private schools would, and common interest would overcome bias of color and lead to more integration than now occurs. The integration would be real, not merely on paper.

The voucher scheme would eliminate the forced busing that a large majority of both blacks and whites object to. Busing would occur, and might indeed increase, but it would be voluntary—just as the busing of children to music and dance classes is today.

The failure of black leaders to espouse vouchers has long puzzled us. Their constituents would benefit most. It would give them control over the schooling of their children, eliminate domination by both the city-wide politicians and, even more important, the entrenched educational bureaucracy. Black leaders frequently send their own children to private schools. Why do they not help others to do the same? Our tentative answer is that vouchers would also free the black man from domination by his own political leaders, who currently see control over schooling as a source of political patronage and power.

However, as the educational opportunities open to the mass of black children have continued to deteriorate, an increasing number of black educators, columnists, and other community leaders have started to support vouchers. The Congress of Racial Equality has made the support of vouchers a major plank in its agenda.

(5) *The economic class issue.* The question that has perhaps divided students of vouchers more than any other is their likely effect on the social and economic class structure. Some have argued that the great value of the public school has been as a melting pot, in which rich and poor, native- and foreign-born, black and white have learned to live together. That image was and is largely true for small communities, but almost entirely false for large cities. There, the public school has fostered residential stratification, by tying the kind and cost of schooling to residential location. It is no accident that most of the country's outstanding public schools are in high-income enclaves.

Most children would still probably attend a neighborhood elementary school under a voucher plan—indeed, perhaps more than now do because the plan would end forced busing. However, because the voucher plan would tend to make residential areas more heterogeneous, the local schools serving any community might well be less homogeneous than they are now. Secondary schools would almost surely be less stratified. Schools defined by common interests—one stressing, say, the arts; another, the sciences; another, foreign languages—would attract students from a wide variety of residential areas. No doubt self-selection would still leave a large class element in the composition of the student bodies, but that element would be less than it is today.

One feature of the voucher plan that has aroused particular concern is the possibility that parents could and would "add on" to the vouchers. If the voucher were for, say, $1,500, a parent could add another $500 to it and send his child to a school charging $2,000 tuition. Some fear that the result might be even wider differences in educational opportunities than now exist because low-income parents would not add to the amount of the voucher while middle-income and upper-income parents would supplement it extensively.

This fear has led several supporters of voucher plans to propose that "add-ons" be prohibited.[16]

Coons and Sugarman write that the

> freedom to add on private dollars makes the Friedman model unacceptable to many, including ourselves. . . . Families unable to add extra dollars would patronize those schools that charged no tuition above the voucher, while the wealthier would be free to distribute themselves among the more expensive schools. What is today merely a personal choice of the wealthy, secured entirely with private funds, would become an invidious privilege assisted by government. . . . This offends a fundamental value commitment—that any choice plan must secure equal family opportunity to attend any participating school.
>
> Even under a choice plan which allowed tuition add-ons, poor families might be better off than they are today. Friedman has argued as much. Nevertheless, however much it improved their education, conscious government finance of economic segregation exceeds our tolerance. If the Friedman scheme were the only politically viable experiment with choice, we would not be enthusiastic.[17]

This view seems to us an example of the kind of egalitarianism discussed in the preceding chapter: letting parents spend money on riotous living but trying to prevent them from spending money on improving the schooling of their children. It is particularly remarkable coming from Coons and Sugarman, who elsewhere say, "A commitment to equality at the deliberate expense of the development of individual children seems to us the final corruption of whatever is good in the egalitarian instinct" [18]—a sentiment with which we heartily agree. In our judgment the very poor would benefit the most from the voucher plan. How can one conceivably justify objecting to a plan, "however much it improved [the] education" of the poor, in order to avoid "government finance of" what the authors call "economic segregation," even if it could be demonstrated to have that effect? And of course, it cannot be demonstrated to have that effect. On the contrary, we are persuaded on the basis of considerable study that it would have precisely the opposite effect—though we must accompany that statement with the qualification that "economic segregation" is so vague a term that it is by no means clear what it means.

The egalitarian religion is so strong that some proponents of restricted vouchers are unwilling to approve even experiments with unrestricted vouchers. Yet to our knowledge, none has ever offered anything other than unsupported assertions to support the fear that an unrestricted voucher system would foster "economic segregation."

This view also seems to us another example of the tendency of intellectuals to denigrate parents who are poor. Even the very poorest can—and do—scrape up a few extra dollars to improve the quality of their children's schooling, although they cannot replace the whole of the present cost of public schooling. We suspect that add-ons would be about as frequent among the poor as among the rest, though perhaps of smaller amounts.

As already noted, our own view is that an unrestricted voucher would be the most effective way to reform an educational system that now helps to shape a life of misery, poverty, and crime for many children of the inner city; that it would undermine the foundations of much of such economic segregation as exists today. We cannot present the full basis for our belief here. But perhaps we can render our view plausible by simply recalling another facet of an earlier judgment: is there any category of goods and services—other than protection against crime—the availability of which currently differs more widely among economic groups than the quality of schooling? Are the supermarkets available to different economic groups anything like so divergent in quality as the schools? Vouchers would improve the quality of the schooling available to the rich hardly at all; to the middle class, moderately; to the lower-income class, enormously. Surely the benefit to the poor more than compensates for the fact that some rich or middle-income parents would avoid paying twice for schooling their children.

(6) *Doubt about new schools.* Is this not all a pipe dream? Private schools now are almost all either parochial schools or elite academies. Will the effect of the voucher plan simply be to subsidize these, while leaving the bulk of the slum dwellers in inferior public schools? What reason is there to suppose that alternatives will really arise?

The reason is that a market would develop where it does not exist today. Cities, states, and the federal government today spend close to $100 billion a year on elementary and secondary schools. That sum is a third larger than the total amount spent annually in restaurants and bars for food and liquor. The smaller sum surely provides an ample variety of restaurants and bars for people in every class and place. The larger sum, or even a fraction of it, would provide an ample variety of schools.

It would open a vast market that could attract many entrants, both from public schools and from other occupations. In the course of talking to various groups about vouchers, we have been impressed by the number of persons who said something like, "I have always wanted to teach [or run a school] but I couldn't stand the educational bureaucracy, red tape, and general ossification of the public schools. Under your plan, I'd like to try my hand at starting a school."

Many of the new schools would be established by nonprofit groups. Others would be established for profit. There is no way of predicting the ultimate composition of the school industry. That would be determined by competition. The one prediction that can be made is that only those schools that satisfy their customers will survive—just as only those restaurants and bars that satisfy their customers survive. Competition would see to that.

(7) *The impact on public schools.* It is essential to separate the rhetoric of the school bureaucracy from the real problems that would be raised. The National Education Association and the American Federation of Teachers claim that vouchers would destroy the public school system, which, according to them, has been the foundation and cornerstone of our democracy. Their claims are never accompanied by any evidence that the public school system today achieves the results claimed for it—whatever may have been true in earlier times. Nor do the spokesmen for these organizations ever explain why, if the public school system is doing such a splendid job, it needs to fear competition from nongovernmental, competitive schools or, if it isn't, why anyone should object to its "destruction."

The threat to public schools arises from their defects, not their accomplishments. In small, closely knit communities where public schools, particularly elementary schools, are now reasonably satisfactory, not even the most comprehensive voucher plan would have much effect. The public schools would remain dominant, perhaps somewhat improved by the threat of potential competition. But elsewhere, and particularly in the urban slums where the public schools are doing such a poor job, most parents would undoubtedly try to send their children to nonpublic schools.

That would raise some transitional difficulties. The parents who are most concerned about their children's welfare are likely to be the first to transfer their children. Even if their children are no smarter than those who remain, they will be more highly motivated to learn and will have more favorable home backgrounds. The possibility exists that some public schools would be left with "the dregs," becoming even poorer in quality than they are now.

As the private market took over, the quality of all schooling would rise so much that even the worst, while it might be *relatively* lower on the scale, would be better in *absolute* quality. And as Harlem Prep and similar experiments have demonstrated, many pupils who are among "the dregs" would perform well in schools that evoked their enthusiasm instead of hostility or apathy.

As Adam Smith put it two centuries ago,

> No discipline is ever requisite to force attendance upon lectures which are really worth the attending. . . . Force and restraint may, no doubt, be in some degree requisite in order to oblige children . . . to attend to those parts of education which it is thought necessary for them to acquire during that early period of life; but after twelve or thirteen years of age, provided the master does his duty, force or restraint can scarce ever be necessary to carry on any part of education. . . .
> Those parts of education, it is to be observed, for the teaching of which there are no public institutions, are generally the best taught.[19]

THE OBSTACLES TO A VOUCHER PLAN

Since we first proposed the voucher plan a quarter-century ago as a practical solution to the defects of the public school system, support has grown. A number of national organizations favor it today.[20] Since 1968 the Federal Office of Economic Opportunity and then the Federal Institute of Education encouraged and financed studies of voucher plans and offered to help finance experimental voucher plans. In 1978 a constitutional amendment was on the ballot in Michigan to mandate a voucher plan. In 1979 a movement was under way in California to qualify a constitutional amendment mandating a voucher plan for the 1980 ballot. A nonprofit institute has recently been established to explore educational vouchers.[21] At the federal level, bills providing for a limited credit against taxes for tuition paid to nonpublic schools have several times come close to passing. While they are not a voucher plan proper, they are a partial variant, partial both because of the limit to the size of the credit and because of the difficulty of including persons with no or low tax liability.

The perceived self-interest of the educational bureaucracy is the key obstacle to the introduction of market competition in schooling. This interest group, which, as Professor Edwin G. West demonstrated, played a key role in the establishment of public schooling in both the United States and Great Britain, has adamantly opposed every attempt to study, explore, or experiment with voucher plans.

Kenneth B. Clark, a black educator and psychologist, summed up the attitude of the school bureaucracy:

> . . . it does not seem likely that the changes necessary for increased efficiency of our urban public schools will come about because they should. . . . What is most important in understanding the ability of the educational establishment to resist change is the fact that public school systems are protected public monopolies with only minimal competition from private and parochial schools. Few critics of the American urban public schools—even severe ones such as myself—dare to question the givens of the present organization of public education. . . . Nor dare the critics question the relevance of the criteria and standards for selecting superintendents, principals, and teachers, or the relevance of all of these to the objectives of public education—producing a literate and informed public to carry on the business of democracy—and to the goal of producing human beings with social sensitivity and dignity and creativity and a respect for the humanity of others.
>
> A monopoly need not genuinely concern itself with these matters. As long as local school systems can be assured of state aid and increasing federal aid without the accountability which inevitably comes with aggressive competition, it would be sentimental, wishful thinking to expect any significant increase in the efficiency of our public schools. If there are no alternatives to the present system—short of present private and parochial schools, which are approaching their limit of expansion—then the possibilities of improvement in public education are limited.[22]

The validity of this assessment was subsequently demonstrated by the reaction of the educational establishment to the federal government's offer to finance experiments in vouchers. Promising initiatives were developed in a considerable number of communities. Only one—at Alum Rock, California—succeeded. It was severely hobbled. The case we know best, from personal experience, was in New Hampshire, where William P. Bittenbender, then chairman of the State Board of Education, was dedicated to conducting an experiment. The conditions seemed excellent, funds were granted by the federal government, detailed plans were drawn up, experimental communities were selected, preliminary agreement from parents and administrators was obtained. When all seemed ready to go, one community after another was persuaded by the local superintendent of schools or other leading figures in the educational establishment to withdraw from the proposed experiment, and the whole venture collapsed.

The Alum Rock experiment was the only one actually to be carried out, and it was hardly a proper test of vouchers. It was limited to a few public schools and allowed no addition to government funds from either parents or others. A number of so-called mini-schools were set up, each with a different curriculum. For three years, parents could choose which their children would attend.[23]

As Don Ayers, who was in charge of the experiment, said, "Probably the most significant thing that happened was that the teachers for the first time had some power and they were able to build the curriculum to fit the needs of the children as they saw it. The state and local school board did not dictate the kind of curriculum that was used in McCollam School. The parents became more involved in the school. They attended more meetings. Also they had a power to pull their child out of that particular mini-school if they chose another mini-school."

Despite the limited scope of that experiment, giving parents greater choice had a major effect on education quality. In terms of test scores, McCollam School went from thirteenth to second place among the schools in its district.

But the experiment is now over, ended by the educational establishment—the same fate that befell Harlem Prep.

The same resistance is present in Great Britain, where an extremely effective group called FEVER (Friends of the Education Voucher Experiment in Representative Regions) have tried for four years to introduce an experiment in a town in the county of Kent, England. The governing authorities have been favorable, but the educational establishment has been adamantly opposed.

The attitude of the professional educators toward vouchers is well expressed by Dennis Gee, headmaster of a school in Ashford, Kent, and secretary of the local teachers' union: "We see this as a barrier between us and the parent—this sticky little piece of paper [i.e., the voucher] in their hand—coming in and under duress—you will do this or else. We make our judgment because we believe it's in the best interest of every Willie and every little Johnny that we've got—and not because someone's going to say 'if you don't do it, we will do that.' It's this sort of philosophy of the marketplace that we object to."

In other words, Mr. Gee objects to giving the customer, in this case the parent, anything to say about the kind of schooling his child gets. Instead, he wants the bureaucrats to decide.

"We are answerable," says Mr. Gee,

> to parents through our governing bodies, through the inspectorate to the Kent County Council, and through Her Majesty's inspectorate to the Secretary of State. These are people, professionals, who are able to make professional judgments.
>
> I'm not sure that parents know what is best educationally for their children. They know what's best for them to eat. They know the best environment they can provide at home. But we've been trained to ascertain the problems of children, to detect their weaknesses, to put right those things that need putting right, and we want to do this freely, with the cooperation of parents and not under undue strains.

Needless to say, at least some parents view things very differently. A local electrical worker and his wife in Kent had to engage in a year-long dispute with the bureaucracy to get their son into the school that they thought was best suited to his needs. Said Maurice Walton,

> As the present system stands, I think we parents have no freedom of choice whatever. They are told what is good for them by the teachers. They are told that the teachers are doing a great job, and they've just got no say at all. If the voucher system were introduced, I think it would bring teachers and parents together—I think closer. The parent that is worried about his child would remove his child from the school that wasn't giving a good service and take it to one that was. . . . If a school was going to crumble because it's got nothing but vandalism, it's generally slack on discipline, and the children aren't learning—well, that's a good thing from my point of view.
>
> I can understand the teachers saying it's a gun at my head, but they've got the same gun at the parents' head at the moment. The parent goes up to the teacher and says, well, I'm not satisfied with what you're doing, and the teacher can say, well tough. You can't take him away, you can't move him, you can't do what you like, so go away and stop bothering me. That can be the attitude of some teachers today, and often is. But now that the positions are being reversed [with vouchers] and the roles are changed, I can only say tough on the teachers. Let them pull their socks up and give us a better deal and let us participate more.

Despite the unrelenting opposition of the educational establishment, we believe that vouchers or their equivalent will be introduced in some form or other soon. We are more optimistic in this area than in welfare because education touches so many of us so deeply. We are willing to make far greater efforts to improve the schooling of our children than to eliminate waste and inequity in the distribution of relief. Discontent with schooling has been rising. So far as we can see, greater parental choice is the only alternative that is available to reduce that discontent. Vouchers keep being rejected and keep emerging with more and more support.

HIGHER EDUCATION: THE PROBLEMS

The problems of higher education in America today, like those in elementary and secondary education, are dual: quality and equity. But in both respects the absence of compulsory attendance alters the problem greatly. No one is required by law to attend an institution of higher education. As a result, students have a wide range of choice about what college or university to attend if they choose to continue their education. A wide range of choice eases the problem of quality, but exacerbates the problem of equity.

Quality. Since no person attends a college or university against his will (or perhaps his parents'), no institution can exist that does not meet, at least to a minimal extent, the demands of its students.

There remains a very different problem. At government institutions at which tuition fees are low, students are second-class customers. They are objects of charity partly supported at the expense of the taxpayer. This feature affects students, faculty, and administrators.

Low tuition fees mean that while city or state colleges and universities attract many serious students interested in getting an education, they also attract many young men and women who come because fees are low, residential housing and food are subsidized, and above all, many other young people are there. For them, college is a pleasant interlude between high school and going to work. Attending classes, taking examinations, getting passing grades—these are the price they are paying for the other advantages, *not* the primary reason they are at school. One result is a high dropout rate. For example, at the University of California in Los Angeles, one of the best regarded state universities in the country, only about half of those who enroll complete the undergraduate course—and this is a high completion rate for government institutions of higher education. Some who drop out transfer to other institutions, but that alters the picture only in detail.

Another result is an atmosphere in the classroom that is often depressing rather than inspiring. Of course, the situation is by no means uniform. Students can choose courses and teachers according to their interest. In every school, serious students and teachers find a way to get together and to achieve their objectives. But again, that is only a minor offset to the waste of students' time and taxpayers' money.

There are good teachers in city and state colleges and universities as well as interested students. But the rewards for faculty and administrators at the prestigious government institutions are not for good undergraduate teaching. Faculty members advance as a result of research and publication; administrators advance by attracting larger appropriations from the state legislature. As a result, even the most famous state universities—the University of California at Los Angeles or at Berkeley, the University of Wisconsin, or the University of Michigan—are not noted for undergraduate teaching. Their reputation is for graduate work, research, and athletic teams—that is where the payoffs are.

The situation is very different at private institutions. Students at such institutions pay high fees that cover much if not most of the cost of their schooling. The money comes from parents, from the students' own earnings, from loans, or from scholarship assistance. The important thing is that the students are the primary customers; they are paying for what they get, and they want to get their money's worth.

The college is selling schooling and the students are buying schooling. As in most private markets, both sides have a strong incentive to serve one another. If the college doesn't provide the kind of schooling its students want, they can go elsewhere. The students want to get full value for their money. As one undergraduate at Dartmouth College, a prestigious private college, remarked, "When you see each lecture costing thirty-five dollars and you think of the other things you can be doing with the thirty-five dollars, you're making very sure that you're going to go to that lecture."

One result is that the fraction of students who enroll at private institutions who complete the undergraduate course is far higher than at government institutions—95 percent at Dartmouth compared to 50 percent at UCLA. The Dartmouth percentage is probably high for private institutions, as the UCLA percentage is for government institutions, but that difference is not untypical.

In one respect this picture of private colleges and universities is oversimplified. In addition to schooling, they produce and sell two other products: monuments and research. Private individuals and foundations have donated most of the buildings and facilities at private colleges and universities, and have endowed professorships and scholarships. Much of the research is financed out of income from endowments or out of special grants from the federal government or other sources for particular purposes. The donors have contributed out of a desire to promote something they regard as desirable. In addition, named buildings, professorships, and scholarships also memorialize an individual, which is why we refer to them as monuments.

The combination of the selling of schooling and monuments exemplifies the much underappreciated ingenuity of voluntary cooperation through the market in harnessing self-interest to broader social objectives. Henry M. Levin, discussing the financing of higher education, writes, "[I]t is doubtful whether the market would support a Classics department or many of the teaching programs in the arts and humanities that promote knowledge and cultural outcomes which are believed widely to affect the general quality of life in our society. The only way these activities would be sustained is by direct social subsidies," by which he means government grants.[24] Mr. Levin is clearly wrong. The market—broadly interpreted—*has* supported social activities in private institutions. And it is precisely because they provide general benefits to society, rather than serving the immediate self-interest of the providers of funds, that they are attractive to donors. Suppose Mrs. X wants to honor her husband, Mr. X. Would she, or anyone else, regard it as much of an honor to have the ABC Manufacturing enterprise (which may be Mr. X's real monument and contribution to social welfare) name a newly built factory for him? On the other hand, if Mrs. X finances a library or other building named for Mr. X at a university, or a named professorship or scholarship, that will be regarded as a real tribute to Mr. X. It will be so regarded precisely because it renders a public service.

Students participate in the joint venture of producing teaching, monuments, and research in two ways. They are customers, but they are also employees. By facilitating the sale of monuments and research, they contribute to the funds available for teaching, thereby earning, as it were, part of their way. This is another example of how complex and subtle are the ways and potentialities of voluntary cooperation.

Many nominally government institutions of higher learning are in fact mixed. They charge tuition and so sell schooling to students. They accept gifts for buildings and the like and so sell monuments. They accept contracts from government agencies or from private enterprises to engage in research. Many state universities have large private endowments—the University of California at Berkeley, the University of Michigan, the University of Wisconsin, to name only a few. Our impression is that the educational performance of the institution has in general been more satisfactory, the larger the role of the market.

Equity. Two justifications are generally offered for using tax money to finance higher education. One, suggested above by Mr. Levin, is that higher education yields "social benefits" over and above the benefits that accrue to the students themselves; the second is that government finance is needed to promote "equal educational opportunity."

(i) *Social benefits.* When we first started writing about higher education, we had a good deal of sympathy for the first justification. We no longer do. In the interim we have tried to induce the people who make this argument to be specific about the alleged social benefits. The answer is almost always simply bad economics. We are told that the nation benefits by having more highly skilled and trained people, that investment in providing such skills is essential for economic growth, that more trained people raise the productivity of the rest of us. These statements are correct. But none is a valid reason for subsidizing higher education. Each statement would be equally correct if made about physical capital (i.e., machines, factory buildings, etc.), yet hardly anyone would conclude that tax money should be used to subsidize the capital investment of General Motors or General Electric. If higher education improves the economic productivity of individuals, they can capture that improvement through higher earnings, so they have a private incentive to get the training. Adam Smith's invisible hand makes their private interest serve the social interest. It is against the social interest to change their private interest by subsidizing schooling. The extra students— those who will only go to college if it is subsidized—are precisely the ones who judge that the benefits they receive are less than the costs. Otherwise they would be willing to pay the costs themselves.

Occasionally the answer is good economics but is supported more by assertion than by evidence. The most recent example is in the reports of a special Commission on Higher Education established by the Carnegie Foundation. In one of its final reports, *Higher Education: Who Pays? Who Benefits? Who Should Pay?*, the commission summarizes the supposed "social benefits." Its list contains the invalid economic arguments discussed in the preceding paragraph—that is, it treats benefits accruing to the persons who get the education as if they were benefits to third parties. But its list also includes some alleged advantages that, if they did occur, would accrue to persons other than those who receive the education, and therefore might justify a subsidy: "general advancement of knowledge . . . ; greater political effectiveness of a democratic society . . . ; greater social effectiveness of society through the resultant better understanding and mutual tolerance among individuals and groups; the more effective preservation and extension of the cultural heritage." [25]

The Carnegie Commission is almost unique in at least paying

some lip service to possible "negative results of higher education" —giving as examples, however, only "the individual frustrations resulting from the current surplus of Ph.D.'s (which is not a social but an individual effect) and the public unhappiness with past outbreaks of campus disruption." [26] Note how selective and biased are the lists of benefits and "negative results." In countries like India, a class of university graduates who cannot find employment they regard as suited to their education has been a source of great social unrest and political instability. In the United States "public unhappiness" was hardly the only, or even the major, negative effect of "campus disruption." Far more important were the adverse effects on the governance of the universities, on the "political effectiveness of a democratic society," on the "social effectiveness of society through . . . better understanding and mutual tolerance"—all cited by the commission, without qualification, as social benefits of higher education.

The report is unique also in recognizing that "without any public subsidy, some of the social benefits of higher education would come as *side effects* of privately financed education in any case." [27] But here again this is simply lip service. Although the commission sponsored numerous and expensive special studies, it did not undertake any serious attempt to identify the alleged social effects in such a way as to permit even a rough quantitative estimate of their importance or of the extent to which they could be achieved without public subsidy. As a result, it offered no evidence that social effects are on balance positive or negative, let alone that any net positive effects are sufficiently large to justify the many billions of dollars of taxpayers' money being spent on higher education.

The commission contented itself with concluding that "no precise—or even imprecise—methods exist to assess the individual and societal benefits as against the private and public costs." But that did not prevent it from recommending firmly and unambiguously an increase in the already massive government subsidization of higher education.

In our judgment this is special pleading, pure and simple. The Carnegie Commission was headed by Clark Kerr, former Chancellor and President of the University of California, Berkeley. Of the eighteen members of the commission, including Kerr, nine either were or had been heads of higher educational institutions, and five others were professionally associated with institutions of higher education. The remaining four had all served on the board of trustees or regents of universities. The academic community has no difficulty recognizing and sneering at special pleading when businessmen march to Washington under the banner of free enterprise to demand tariffs, quotas, and other special benefits. What would the academic world say about a steel industry commission, fourteen of whose eighteen members were from the steel industry, which recommended a major expansion in government subsidies to the steel industry? Yet we have heard nothing from the academic world about the comparable recommendation of the Carnegie Commission.

(ii) *Equal educational opportunity.* The promotion of "equal educational opportunity" is the major justification that is generally offered for using tax money to finance higher education. In the words of the Carnegie Commission, "We have favored . . . [a] larger public . . . share of monetary outlays for education on a temporary basis in order to make possible greater equality of educational opportunity." [28] In the words of the parent Carnegie Foundation, "Higher education is . . . a major avenue to greater equality of opportunity, increasingly favored by those whose origins are in low-income families and by those who are women and members of minority groups." [29]

The objective is admirable. The statement of fact is correct. But there is a missing link between the one and the other. Has the objective been promoted or retarded by government subsidy? Has higher education been a "major avenue to greater equality of opportunity" because of or despite government subsidy?

One simple statistic from the Carnegie Commission's own report illustrates the problem of interpretation: 20 percent of college students from families with incomes below $5,000 in 1971 attended private institutions; 17 percent from families with incomes between $5,000 and $10,000; 25 percent from families with incomes over $10,000. In other words, the private institutions provided more opportunity for young men and women at the very bottom as well as the top of the income scale than did the government institutions. [30]

And this is just the tip of the iceberg. Persons from middle-

and upper-income families are two or three times as likely to attend college as persons from lower-income groups, and they go to school for more years at the more expensive institutions (four-year colleges and universities rather than two-year junior colleges). As a result, student from higher-income families benefit the most from the subsidies.[31]

Some persons from poor families do benefit from the government subsidy. In general, they are the ones among the poor who are better off. They have human qualities and skills that will enable them to profit from higher education, skills that would also have enabled them to earn a higher income without a college education. In any event, they are destined to be among the better off in the community.

Two detailed studies, one for Florida, one for California, underline the extent to which government spending on higher education transfers income from low- to high-income groups.

The Florida study compared the total benefits persons in each of four income classes received in 1967–68 from government expenditures on higher education with the costs they incurred in the form of taxes. Only the top income class got a net gain; it got back 60 percent more than it paid. The bottom two classes paid 40 percent more than they got back, the middle class nearly 20 percent more.[32]

The California study, for 1964, is just as striking, though the key results are presented somewhat differently, in terms of families with and without children in California public higher education. Families with children in public higher education received a net benefit varying from 1.5 percent to 6.6 percent of their average income, the largest benefit going to those who had children at the University of California and who also had the highest average income. Families without children in public higher education had the lowest average income and incurred a net cost of 8.2 percent of their income.[33]

The facts are not in dispute. Even the Carnegie Commission admits the perverse redistributive effect of government expenditures on higher education—although one must read their reports with great care, and indeed between the lines, to spot the admission in such comments as, "This 'middle class' generally . . .

does quite well in the proportion of public subsidies that it receives. Greater equity can be achieved through a reasonable redistribution of subsidies." [34] Its major solution is more of the same: still greater government spending on higher education. We know of no government program that seems to us so inequitable in its effects, so clear an example of Director's Law, as the financing of higher education. In this area those of us who are in the middle- and upper-income classes have conned the poor into subsidizing us on the grand scale—yet we not only have no decent shame, we boast to the treetops of our selflessness and public-spiritedness.

HIGHER EDUCATION: THE SOLUTION

It is eminently desirable that every young man and woman, regardless of his or her parents' income, social position, residence, or race, have the opportunity to get higher education—*provided that he or she is willing to pay for it either currently or out of the higher income the schooling will enable him or her to earn.* There is a strong case for providing loan funds sufficient to assure opportunity to all. There is a strong case for disseminating information about the availability of such funds and for urging the less privileged to take advantage of the opportunity. There is no case for subsidizing persons who get higher education at the expense of those who do not. Insofar as governments operate institutions of higher education, they should charge students fees corresponding to the full cost of the educational and other services they provide to them.

However desirable it may be to eliminate taxpayer subsidization of higher education, that does not currently seem politically feasible. Accordingly, we shall supplement our discussion of an alternative to government finance with a less radical reform— a voucher plan for higher education.

Alternative to government finance. Fixed-money loans to finance higher schooling have the defect that there is wide diversity in the earnings of college graduates. Some will do very well. Paying back a fixed-dollar loan would be no great problem for them. Others will end with only modest incomes. They would find a fixed debt a heavy burden. Expenditure on education is a capital investment in a risky enterprise, as it were, like investment in a newly formed small business. The most satisfactory method of financing such enterprises is not through a fixed-dollar loan but through equity investment—"buying" a share in the enterprise and receiving as a return a share of the profits.

For education, the counterpart would be to "buy" a share in an individual's earning prospects, to advance him the funds needed to finance his training on condition that he agree to pay the investor a specified fraction of his future earnings. In this way an investor could recoup more than his initial investment from relatively successful individuals, which would compensate for the failure to do so from the unsuccessful. Though there seems no legal obstacle to private contracts on this basis, they have not become common, primarily, we conjecture, because of the difficuly and costs of enforcing them over the long period involved.

A quarter-century ago (1955), one of us published a plan for "equity" financing of higher education through a government body that

> could offer to finance or help finance the training of any individual who could meet minimum quality standards. It would make available a limited sum per year for a specified number of years, provided the funds were spent on securing training at a recognized institution. The individual in return would agree to pay to the government in each future year a specified percentage of his earnings in excess of a specified sum for each $1,000 that he received from the government. This payment could easily be combined with the payment of income tax and so involve a minimum of additional administrative expense. The base sum should be set equal to estimated average earnings without the specialized training; the fraction of earnings paid should be calculated so as to make the whole project self-financing. In this way, the individuals who received the training would in effect bear the whole cost. The amount invested could then be determined by individual choice.[35]

More recently (1967), a panel appointed by President Johnson and headed by Professor Jerrold R. Zacharias of MIT recommended the adoption of a specific version of this plan under the appealing title "Educational Opportunity Bank" and made an extensive and detailed study of its feasibility and of the terms that would be required in order for it to be self-supporting.[36] No reader of this book will be surprised to learn that the proposal was met by a blast from the Association of State Universities and Land Grant Colleges—a fine example of what Adam Smith referred to as "the passionate confidence of interested falsehood."[37]

In 1970, as recommendation 13 out of thirteen recommendations for the financing of higher education, the Carnegie Commission proposed the establishment of a National Student Loan Bank that would make long-term loans with repayment partly contingent upon current earnings. "Unlike the Educational Opportunity Bank," says the commission, ". . . we see the National Student Loan Bank as a means of providing supplementary funding for students, not as a way of financing total educational costs."[38]

More recently still, some universities, including Yale University, have considered or adopted contingent-repayment plans administered by the university itself. So a spark of life remains.

A voucher plan for higher education. Insofar as any tax money is spent to subsidize higher education, the least bad way to do so is by a voucher arrangement like that discussed earlier for elementary and secondary schools.

Have all government schools charge fees covering the full cost of the educational services they provide and so compete on equal terms with nongovernment schools. Divide the total amount of taxes to be spent annually on higher education by the number of students it is desired to subsidize per year. Give that number of students vouchers equal to the resulting sum. Permit the vouchers to be used at any educational institution of the student's choice, provided only that the schooling is of a kind that it is desired to subsidize. If the number of students requesting vouchers is greater than the number available, ration the vouchers by whatever criteria the community finds most acceptable: competitive examinations, athletic ability, family income, or any of myriad other possible standards. The resulting system would follow in broad outline the GI bills providing for the education of veterans, except that the GI bills were open-ended; their benefits were available to all veterans.

As we wrote when we first proposed this plan:

> The adoption of such arrangements would make for more effective competition among various types of schools and for a more efficient utilization of their resources. It would eliminate the pressure for direct government assistance to private colleges and universities and thus preserve their full independence and diversity at the same time as it enabled them to grow relative to state institutions. It might also have the ancillary advantage of causing scrutiny of the purposes for which subsidies are granted. The subsidization of institutions rather than of people has led to an indiscriminate subsidization of all activities appropriate for such institutions, rather than of the activities appropriate for the state to subsidize. Even cursory examination suggests that while the two classes of activities overlap, they are far from identical.
>
> The equity argument for the alternative [voucher] arrangement is . . . clear. . . . The state of Ohio, for example, says to its citizens: "If you have a youngster who wants to go to college, we shall automatically give him or her a sizable four-year scholarship, provided that he or she can satisfy rather minimal education requirements, and provided further that he or she is smart enough to choose to go to the University of Ohio [or some other state-supported institution]. If your youngster wants to go, or you want him or her to go, to Oberlin College, or Western Reserve University, let alone to Yale, Harvard, Northwestern, Beloit, or the University of Chicago, not a penny for him." How can such a program be justified? Would it not be far more equitable, and promote a higher standard of scholarship, to devote such money as the state of Ohio wished to spend on higher education to scholarships tenable at any college or university and to require the University of Ohio to compete on equal terms with other colleges and universities? [39]

Since we first made this proposal, a number of states have adopted a limited program going partway in its direction by giving scholarships tenable at private colleges and universities, though only those in the state in question. On the other hand, an excellent program of Regents scholarships in New York State, very much in the same spirit, was emasculated by Governor Nelson Rockefeller's grandiose plans for a State University of New York modeled after the University of California.

Another important development in higher education has been a major expansion in the federal government's involvement in financing, and even more in regulating both government and nongovernment institutions. The intervention has in large measure been part of the greatly expanded federal activity to foster so-called "affirmative action," in the name of greater civil rights. This intervention has aroused great concern among faculty and administrators at colleges and universities, and much opposition by them to the activities of federal bureaucrats.

The whole episode would be a matter of poetic justice if it were not so serious for the future of higher education. The academic community has been in the forefront of the proponents of such intervention—when directed at other segments of society. They have discovered the defects of intervention—its costliness, its interference with the primary mission of the institutions, and its counterproductiveness in its own terms—only when these measures were directed at them. They have now become the victims both of their own earlier professions of faith and of their self-interest in continuing to feed at the federal trough.

CONCLUSION

In line with common practice, we have used "education" and "schooling" as synonymous. But the identification of the two terms is another case of using persuasive terminology. In a more careful use of the terms, not all "schooling" is "education," and not all "education" is "schooling." Many highly schooled people are uneducated, and many highly "educated" people are unschooled.

Alexander Hamilton was one of the most truly "educated," literate, and scholarly of our founding fathers, yet he had only three or four years of formal schooling. Examples could be multiplied manyfold, and no doubt every reader knows highly schooled people whom he regards as uneducated and unschooled people whom he considers learned.

We believe that the growing role that government has played in financing and administering schooling has led not only to enormous waste of taxpayers' money but also to a far poorer educational system than would have developed had voluntary cooperation continued to play a larger role.

Few institutions in our society are in a more unsatisfactory state than schools. Few generate more discontent or can do more to undermine our liberty. The educational establishment is up in arms in defense of its existing powers and privileges. It is supported by many public-spirited citizens who share a collectivist outlook. But it is also under attack. Declining test scores throughout the country; increasing problems of crime, violence, and disorder at urban schools; opposition on the part of the overwhelming majority of both whites and blacks to compulsory busing; restiveness on the part of many college and university teachers and administrators under the heavy hand of HEW bureaucrats—all this is producing a backlash against the trend toward centralization, bureaucratization, and socialization of schooling.

We have tried in this chapter to outline a number of constructive suggestions: the introduction of a voucher system for elementary and secondary education that would give parents at all income levels freedom to choose the schools their children attend; a contingent-loan financing system for higher education to combine equality of opportunity with the elimination of the present scandalous imposition of taxes on the poor to pay for the higher education of the well-to-do; or, alternatively, a voucher plan for higher education that would both improve the quality of institutions of higher education and promote greater equity in the distribution of such taxpayer funds as are used to subsidize higher education.

These proposals are visionary but they are not impracticable. The obstacles are in the strength of vested interests and prejudices, not in the feasibility of administering the proposals. There are forerunners, comparable programs in operation in this country and elsewhere on a smaller scale. There is public support for them.

We shall not achieve them at once. But insofar as we make progress toward them—or alternative programs directed at the same objective—we can strengthen the foundations of our freedom and give fuller meaning to equality of educational opportunity.

CHAPTER 7

Who Protects
the Consumer?

"It is not from the benevolence of the butcher, the brewer, or the baker, that we expect our dinner, but from their regard to their own interest. We address ourselves, not to their humanity but to their self-love, and never talk to them of our own necessities but of their advantages. Nobody but a beggar chuses to depend chiefly upon the benevolence of his fellow citizens."
—Adam Smith, *The Wealth of Nations*, vol. I, p. 16

We cannot indeed depend on benevolence for our dinner—but can we depend wholly on Adam Smith's invisible hand? A long line of economists, philosophers, reformers, and social critics have said no. Self-love will lead sellers to deceive their customers. They will take advantage of their customers' innocence and ignorance to overcharge them and pass off on them shoddy products. They will cajole customers to buy goods they do not want. In addition, the critics have pointed out, if you leave it to the market, the outcome may affect people other than those directly involved. It may affect the air we breathe, the water we drink, the safety of the foods we eat. The market must, it is said, be supplemented by other arrangements in order to protect the consumer from himself and from avaricious sellers, and to protect all of us from the spillover neighborhood effects of market transactions.

These criticisms of the invisible hand are valid, as we noted in Chapter 1. The question is whether the arrangements that have been recommended or adopted to meet them, to supplement the market, are well devised for that purpose, or whether, as so often happens, the cure may not be worse than the disease.

This question is particularly relevant today. A movement launched less than two decades ago by a series of events—the publication of Rachel Carson's *Silent Spring*, Senator Estes Kefauver's investigation of the drug industry, and Ralph Nader's attack on the General Motors Corvair as "unsafe at any speed"— has led to a major change in both the extent and the character of government involvement in the marketplace—in the name of protecting the consumer.

From the Army Corps of Engineers in 1824 to the Interstate Commerce Commission in 1887 to the Federal Railroad Administration in 1966, the agencies established by the federal government to regulate or supervise economic activity varied in scope, importance, and purpose, but almost all dealt with a single industry and had well-defined powers with respect to that industry. From at least the ICC on, protection of the consumer—primarily his pocketbook—was one objective proclaimed by the reformers.

The pace of intervention quickened greatly after the New Deal —half of the thirty-two agencies in existence in 1966 were created after FDR's election in 1932. Yet intervention remained fairly moderate and continued in the single-industry mold. The *Federal Register*, established in 1936 to record all the regulations, hearings, and other matters connected with the regulatory agencies, grew, at first rather slowly, then more rapidly. Three volumes, containing 2,599 pages and taking six inches of shelf space, sufficed for 1936; twelve volumes, containing 10,528 pages and taking twenty-six inches of shelf space, for 1956; and thirteen volumes, containing 16,850 pages and taking thirty-six inches of shelf space, for 1966.

Then a veritable explosion in government regulatory activity occurred. No fewer than twenty-one new agencies were established in the next decade. Instead of being concerned with specific industries, they covered the waterfront: the environment, the production and distribution of energy, product safety, occupational safety, and so on. In addition to concern with the consumer's pocketbook, with protecting him from exploitation by sellers, recent agencies are primarily concerned with things like the consumer's safety and well-being, with protecting him not only from sellers but also from himself.[1]

Government expenditures on both older and newer agencies skyrocketed—from less than $1 billion in 1970 to roughly $5 billion estimated for 1979. Prices in general roughly doubled, but these expenditures more than quintupled. The number of government bureaucrats employed in regulatory activities tripled, going from 28,000 in 1970 to 81,000 in 1979; the number of pages in the *Federal Register*, from 17,660 in 1970 to 36,487 in 1978, taking 127 inches of shelf space—a veritable ten-foot shelf.

During the same decade, economic growth in the United States slowed drastically. From 1949 to 1969, output per man-hour of all persons employed in private business—a simple and comprehensive measure of productivity—rose more than 3 percent a year; in the next decade, less than half as fast; and by the end of the decade productivity was actually declining.

Why link these two developments? One has to do with assuring our safety, protecting our health, preserving clean air and water; the other, with how effectively we organize our economy. Why should these two good things conflict?

The answer is that whatever the announced objectives, all of the movements in the past two decades—the consumer movement, the ecology movement, the back-to-the-land movement, the hippie movement, the organic-food movement, the protect-the-wilderness movement, the zero-population-growth movement, the "small is beautiful" movement, the antinuclear movement—have had one thing in common. All have been antigrowth. They have been opposed to new developments, to industrial innovation, to the increased use of natural resources. Agencies established in response to these movements have imposed heavy costs on industry after industry to meet increasingly detailed and extensive government requirements. They have prevented some products from being produced or sold; they have required capital to be invested for nonproductive purposes in ways specified by government bureaucrats.

The results have been far-reaching and threaten to be even more so. As Edward Teller, the great nuclear physicist, once put it, "It took us eighteen months to build the first nuclear power generator; it now takes twelve years; that's progress." The direct cost of regulation to the taxpayer is the least part of its total cost. The $5 billion a year spent by the government is swamped by the costs to industry and consumer of complying with the regulations. Conservative estimates put that cost at something like $100 billion a year. And that doesn't count the cost to the consumer of restricted choice and higher prices for the products that are available.

This revolution in the role of government has been accompanied, and largely produced, by an achievement in public persuasion that must have few rivals. Ask yourself what products are currently least satisfactory and have shown the least improvement over time. Postal service, elementary and secondary schooling, railroad passenger transport would surely be high on the list. Ask yourself which products are most satisfactory and have improved the most. Household appliances, television and radio sets, hi-fi equipment, computers, and, we would add, supermarkets and shopping centers would surely come high on that list.

The shoddy products are all produced by government or government-regulated industries. The outstanding products are all produced by private enterprise with little or no government involvement. Yet the public—or a large part of it—has been persuaded that private enterprises produce shoddy products, that we need ever vigilant government employees to keep business from foisting off unsafe, meretricious products at outrageous prices on ignorant, unsuspecting, vulnerable customers. That public relations campaign has succeeded so well that we are in the process of turning over to the kind of people who bring us our postal service the far more critical task of producing and distributing energy.

Ralph Nader's attack on the Corvair, the most dramatic single episode in the campaign to discredit the products of private industry, exemplifies not only the effectiveness of that campaign but also how misleading it has been. Some ten years after Nader castigated the Corvair as unsafe at any speed, one of the agencies that was set up in response to the subsequent public outcry finally got around to testing the Corvair that started the whole thing. They spent a year and a half comparing the performance of the Corvair with the performance of other comparable vehicles, and they concluded, "The 1960–63 Corvair compared favorably with the other contemporary vehicles used in the tests." [2] Nowadays Corvair fan clubs exist throughout the country. Corvairs have become collectors' items. But to most people, even the well informed, the Corvair is still "unsafe at any speed."

The railroad industry and the automobile industry offer an excellent illustration of the difference between a governmentally regulated industry protected from competition and a private industry subjected to the full rigors of competition. Both industries serve the same market and ultimately provide the same service, transportation. One industry is backward and inefficient and displays little innovation. The major exception was the replacement of the steam engine by the diesel. The freight cars being pulled by the diesels today are hardly distinguishable from those that were pulled by the steam engines of an earlier era. Passenger service is slower and less satisfactory today than it was fifty years ago. The railroads are losing money and are in the process of being taken over by the government. The automobile industry, on the other hand, spurred by competition from home and abroad and free to innovate, has made tremendous strides, introducing one innovation after another, so that the cars of fifty years ago are museum pieces. The consumers have benefited—and so have the workers and stockholders in the automobile industry. Impressive —and tragic, because the automobile industry is now rapidly being converted into a governmentally regulated industry. We can see the developments that hobbled railroads occurring before our very eyes to automobiles.

Government intervention in the marketplace is subject to laws of its own, not legislated laws, but scientific laws. It obeys forces and goes in directions that may have little relationship to the intentions or desires of its initiators or supporters. We have already examined this process in connection with welfare activity. It is present equally when government intervenes in the marketplace, whether to protect consumers against high prices or shoddy goods, to promote their safety, or to preserve the environment. Every act of intervention establishes positions of power. How that power will be used and for what purposes depends far more on the people who are in the best position to get control of that power and what their purposes are than on the aims and objectives of the initial sponsors of the intervention.

The Interstate Commerce Commission, dating from 1887, was the first agency established largely through a political crusade led by self-proclaimed representatives of the consumer—the Ralph Naders of the day. It has gone through several life cycles and has been exhaustively studied and analyzed. It provides an excellent example to illustrate the natural history of government intervention in the marketplace.

The Food and Drug Administration, initially established in 1906 in response to the outcry that followed Upton Sinclair's novel *The Jungle*, which exposed unsanitary conditions in the Chicago slaughtering and meat-packing houses, has also gone through several life cycles. Aside from its intrinsic interest, it serves as something of a bridge between the earlier specific-industry type of regulation and the more recent functional or cross-industry type of regulation because of the change that occurred in its activities after the 1962 Kefauver amendments.

The Consumer Products Safety Commission, the National Highway Traffic Safety Administration, the Environmental Protection Agency, all exemplify the more recent type of regulatory agency—cutting across industry and relatively unconcerned with the consumer's pocketbook. A full analysis of them is far beyond our scope, but we discuss briefly how they exemplify the same tendencies that are present in ICC and FDA, and the problems they raise for the future.

Though intervention in energy by both state and federal governments is of long standing, there was a quantum jump after the OPEC embargo in 1973 and subsequent quadrupling of the price of crude oil.

If, as we shall argue, we cannot depend on government intervention to protect us as consumers, what can we depend on? What devices does the market develop for that purpose? And how can they be improved?

THE INTERSTATE COMMERCE COMMISSION

The Civil War was followed by an unprecedented expansion of the railroads—symbolized by the driving of the Golden Spike at Promontory Point, Utah, on May 10, 1869, to mark the joining of the Union Pacific and Central Pacific railroads, completing the first transcontinental line. Soon there was a second, third, and even fourth transcontinental route. In 1865 railroads already operated 35,000 miles of track; ten years later, close to 75,000; and by 1885, over 125,000. By 1890 there were more than 1,000 separate railroads. The country was literally crisscrossed with railroads going to every remote hamlet and covering the nation from coast to coast. The miles of track in the United States exceeded that in all the rest of the world combined. Competition was fierce. As a result, freight and passenger rates were low, supposedly the lowest in the world. Railroad men, of course, complained of "cutthroat competition." Every time the economy faltered, in one of its periodic slumps, railroads went bankrupt and were taken over by others or simply went out of business. When the economy revived, another surge of railroad construction followed.

The railroad men of the time tried to improve their position by joining together, forming pools, agreeing to fix rates at profitable levels and to divide the market. To their dismay, the agreements were always breaking down. So long as the rest of the members of a pool kept up their rates, any one member could benefit by cutting his rates and taking business away from the others. Of course, he would not cut rates openly; he would do so in devious ways to keep the other members of the pool in the dark as long as possible. Hence such practices arose as secret rebates to favored shippers and discriminatory pricing between regions or commodities. Sooner or later the price cutting would become known and the pool would collapse.

Competition was fiercest between distant, populous points such as New York and Chicago. Shippers and passengers could choose among a number of alternate routes operated by different railroads and also among the canals that had earlier covered the land. On the other hand, between shorter segments of any one of these routes, for example, between Harrisburg and Pittsburgh, there might be only one railroad. That railroad would have something of a monopoly position, subject only to competition from alternative means of transport, such as canals or rivers. Naturally, it would take full advantage of its monopoly position wherever it could and charge all that the traffic would bear.

One result was that the sum of the fares charged for the short hauls—or even for one short haul—was sometimes larger than the total sum charged for the long haul between the two distant points. Of course, none of the consumers complained about the low prices for the long haul, but they certainly did complain about the higher prices for the short hauls. Similarly, the favored shippers who got rebates in the secret rate-cutting wars did not complain, but those who failed to get rebates were loud in their complaints about "discriminatory pricing."

The railroads were the major enterprises of the day. Highly visible, highly competitive, linked with Wall Street and the financial East, they were a steady source of stories of financial manipulation and skulduggery in high places. They became a natural target, particularly for the farmers of the Middle West. The Grange movement, which arose in the 1870s, attacked the "monopolistic railroads." They were joined by the Greenback party, the Farmers' Alliance, and so on and on, all agitating, frequently with success, at the statehouse for government control of freight rates and practices. The Populist party, through which William Jennings Bryan rose to fame, called not merely for regulation of the railroads but for outright government ownership and operation.[3] The cartoonists of the time had a field day depicting the railroads as octopuses strangling the country and exercising tremendous political influence—which indeed they did.

As the campaign against the railroads mounted, some farsighted railroad men recognized that they could turn it to their advantage, that they could use the federal government to enforce their price-fixing and market-sharing agreements and to protect themselves from state and local governments. They joined the reformers in supporting government regulation. The outcome was the establishment of the Interstate Commerce Commission in 1887.

It took about a decade to get the commission in full operation. By that time the reformers had moved on to their next crusade. The railroads were only one of their concerns. They had achieved their objective, and they had no overpowering interest to lead them to do more than cast an occasional glance at what the ICC was doing. For the railroad men the situation was entirely different. The railroads were their business, their overriding concern. They were prepared to spend twenty-four hours a day on it. And who else had the expertise to staff and run the ICC? They soon learned how to use the commission to their own advantage.

The first commissioner was Thomas Cooley, a lawyer who had represented the railroads for many years. He and his associates sought greater regulatory power from Congress, and that power was granted. As President Cleveland's Attorney General, Richard J. Olney, put it in a letter to railroad tycoon Charles E. Perkins, president of the Burlington & Quincy Railroad, only a half-dozen years after the establishment of the ICC:

> The Commission, as its functions have now been limited by the courts, is, or can be made, of great use to the railroads. It satisfies the popular clamor for a Government supervision of railroads, at the same time that that supervision is almost entirely nominal. Further, the older such a commission gets to be, the more inclined it will be found to take the business and railroad view of things. It thus becomes a sort of barrier between the railroad corporations and the people and a sort of protection against hasty and crude legislation hostile to railroad interests. . . . The part of wisdom is not to destroy the Commission, but to utilize it.[4]

The commission solved the long-haul/short-haul problem. As you will not be surprised to learn, it did so mostly by *raising* the long-haul rates to equal the sum of the short-haul rates. Everybody except the customer was happy.

As time passed, the commission's powers were increased and it came to exercise closer and closer control over every aspect of the railroad business. In addition, power shifted from direct representatives of the railroads to the growing ICC bureaucracy. However, that was no threat to the railroads. Many of the bureaucrats were drawn from the railroad industry, their day-to-day business tended to be with railroad people, and their chief hope of a lucrative future career was with railroads.

The real threat to the railroads arose in the 1920s, when trucks emerged as long-distance haulers. The artificially high freight rates maintained by the ICC for railroads enabled the trucking industry to grow by leaps and bounds. It was unregulated and highly competitive. Anybody with enough capital to buy a truck could go into the business. The principal argument used against the railroads in the campaign for government regulation—that they were monopolies that had to be controlled to keep them from exploiting the public—had no validity whatsoever for trucking.

It would be hard to find an industry that came closer to satisfying the requirements for what the economists call "perfect" competition.

But that did not stop the railroads from agitating to have long-distance trucking brought under the control of the Interstate Commerce Commission. And they succeeded. The Motor Carrier Act of 1935 gave the ICC jurisdiction over truckers—to protect the railroads, not the consumers.

The railroad story was repeated for trucking. It was cartelized, rates were fixed, routes assigned. As the trucking industry grew, the representatives of the truckers came to have more and more influence on the commission and gradually came to replace railroad representatives as the dominant force. The ICC became as much an agency devoted to protecting the trucking industry from the railroads and the nonregulated trucks as to protecting the railroads against the trucks. With it all, there was an overlay of simply protecting its own bureaucracy.

In order to operate as an interstate public carrier, a trucking company must have a certificate of public convenience and necessity issued by the ICC. Out of some 89,000 initial applications for such certificates after the passage of the Motor Carrier Act of 1935, the ICC approved only about 27,000. "Since that time . . . the commission has been very reluctant to grant new competitive authority. Moreover, mergers and failures of existing trucking firms have reduced the number of such firms from over 25,000 in 1939 to 14,648 in 1974. At the same time, the tons shipped by regulated trucks in intercity service have increased from 25.5 million in 1938 to 698.1 million in 1972: a 27-fold increase." [5]

The certificates can be bought and sold. "The growth in traffic, the decline in number of firms, and the discouragement of rate competition by rate bureaus and ICC practices have increased the value of certificates considerably." Thomas Moore estimates that their aggregate value in 1972 was between $2 and $3 billion[6]— a value that corresponds solely to a government-granted monopoly position. It constitutes wealth for the people who own the certificates, but for the society as a whole it is a measure of the loss from government intervention, not a measure of productive capacity. Every study shows that the elimination of ICC regulation of trucking would drastically reduce costs to shippers—Moore estimates by perhaps as much as three-quarters.

A trucking company in Ohio, Dayton Air Freight, offers a specific example. It has an ICC license that gives it exclusive permission to carry freight from Dayton to Detroit. To serve other routes it has had to buy rights from ICC license holders, including one who doesn't own a single truck. It has paid as much as $100,000 a year for the privilege. The owners of the firm have been trying to get their license extended to cover more routes, so far without success.

As one of their customers, Malcolm Richards, put it, "Quite frankly I don't know why the ICC is sitting on its hands doing nothing. This is the third time to my knowledge that we have supported the application of Dayton Air Freight to help us save money, help free enterprise, help the country save energy. . . . It all comes down to the consumer's ultimately going to pay for all this."

One of the owners of Dayton Air Freight, Ted Hacker, adds: "As far as I'm concerned, there is no free enterprise in interstate commerce. It no longer exists in this country. You have to pay the price and you have to pay the price very dearly. And that not only means that we have to pay the price, it means the consumer is paying the price."

But this comment has to be taken with a real grain of salt in light of a comment by another owner, Herschel Wimmer: "I have no argument with the people who already have ICC permits excepting for the fact this is a big country and since the inception of the ICC in 1936, there have been few entrants into the business. They do not allow new entrants to come into the business and compete with those who are already in."

We conjecture that this reflects a reaction we have encountered repeatedly among railroad men and truckers: give us a certificate or grant us a waiver of the rules, yes; abolish the issuance of certificates or the system of government regulation, no. In view of the vested interests that have grown up, that reaction is entirely understandable.

To return to railroads, the ultimate effects of government intervention are not yet over. The increasingly rigid rules prevented railroads from adjusting effectively to the emergence of auto-mobiles, buses, and planes as an alternative to railroads for long-distance passenger traffic. They once again turned to the govern-ment, this time by the nationalization of passenger traffic in the form of Amtrak. The same process is occurring in freight. Much of the railroad freight trackage in the Northeast has in effect been nationalized through the creation of Conrail following the dra-matic bankruptcy of the New York Central Railroad. That is very likely the prospect for the rest of the railroad industry as well.

Air travel repeated the railroad and trucking story. When the Civil Aeronautics Board was established in 1938, it assumed con-trol over nineteen domestic trunk line carriers. Today there are even fewer, despite the enormous growth in air travel, and despite numerous applications for "certificates of public convenience and necessity." The airline story does differ in one important respect. For a variety of reasons—not least the successful price cutting across the Atlantic by Freddie Laker, the enterprising British owner of a major international airline, and the personality and ability of Alfred Kahn, former chairman of the CAB—there has recently been considerable deregulation of air fares, both ad-ministratively and legislatively. This is the first major move in any area away from government control and toward greater freedom. Its dramatic success—lower fares yet higher earnings for the air-lines—has encouraged a movement toward some measure of deregulation of surface transportation. However, powerful forces, particularly in the trucking industry, are organizing opposition to such deregulation, so as yet it is only a faint hope.

One ironic echo of the long-haul/short-haul issue recently arose in the air industry. In this case the discrepancy was the opposite of that in rails—the short-haul fare was the lower. The case occurred in California, which is a large enough state to support several major airlines that fly solely within the state and as a result were not subject to CAB control. Competition on the route between San Francisco and Los Angeles produced an intrastate fare that was much lower than the fare that the CAB permitted interstate lines to charge for the same trip.

The irony is that a complaint was filed before the CAB about the discrepancy in 1971 by Ralph Nader, self-proclaimed de-

fender of the consumer. It so happens that one of Nader's sub-
sidiaries had published an excellent analysis of the ICC, stressing,
among other things, how the long-haul/short-haul discrimination
was resolved. Nader could hardly have been under any illusions
about how the airline case would be resolved. As any student of
regulation would have predicted, the CAB ruling, later upheld by
the Supreme Court, required intrastate companies to raise their
fares to match those permitted by CAB. Fortunately, the ruling
was in abeyance because of legal technicalities and may be rendered irrelevant by the deregulation of air fares.

The ICC illustrates what might be called the natural history of
government intervention. A real or fancied evil leads to demands
to do something about it. A political coalition forms consisting of
sincere, high-minded reformers and equally sincere interested
parties. The incompatible objectives of the members of the coalition (e.g., low prices to consumers and high prices to producers)
are glossed over by fine rhetoric about "the public interest," "fair
competition," and the like. The coalition succeeds in getting Congress (or a state legislature) to pass a law. The preamble to the
law pays lip service to the rhetoric and the body of the law grants
power to government officials to "do something." The high-minded
reformers experience a glow of triumph and turn their attention
to new causes. The interested parties go to work to make sure
that the power is used for their benefit. They generally succeed.
Success breeds its problems, which are met by broadening the
scope of intervention. Bureaucracy takes its toll so that even the
initial special interests no longer benefit. In the end the effects are
precisely the opposite of the objectives of the reformers and generally do not even achieve the objectives of the special interests.
Yet the activity is so firmly established and so many vested interests are connected with it that repeal of the initial legislation is
nearly inconceivable. Instead, new government legislation is called
for to cope with the problems produced by the earlier legislation
and a new cycle begins.

The ICC reveals clearly each of these steps—from the curious
coalition responsible for its establishment to the beginning of a
second cycle by the establishment of Amtrak, whose only excuse
for existence is that it is largely free from ICC regulation and can

therefore do what ICC will not permit the individual railroads to do. The rhetoric, of course, was that the purpose of Amtrak was improved rail passenger transportation. It was supported by railroads because it would permit much then-existing passenger service to be eliminated. The excellent and profitable passenger service of the 1930s had deteriorated and become unprofitable as a result of the competition of the airplane and the private car. Yet ICC would not permit the railroads to curtail the service. Amtrak is now both curtailing it and subsidizing what remains.

If the ICC had never been established and market forces had been permitted to operate, the United States would today have a far more satisfactory transportation system. The railroad industry would be leaner but more efficient as a result of greater technological innovation under the spur of competition and the more rapid adjustment of routes to the changing demands of traffic. Passenger trains might serve fewer communities but the facilities and equipment would be far better than they are now, and the service more convenient and rapid.

Similarly, there would be more trucking firms though there might be fewer trucks because of greater efficiency and less waste in such forms as the empty return trips and roundabout routes that ICC regulations now mandate. Costs would be lower and service better. The reader who has had occasion to use an ICC-licensed company to move his personal belongings will have no difficulty in accepting that judgment. Though we do not speak from personal experience, we suspect that this is also true for commercial shippers.

The whole shape of the transportation industry might be radically different, involving perhaps much greater use of combined modes of transport. One of the few profitable private railroad operations in recent years has been a service transporting people plus their automobiles in the same train. Piggyback operation would doubtless have been introduced much sooner than it was, and many other combinations might have emerged.

A major argument for letting market forces work is the very difficulty of imagining what the outcome would be. The one thing that is certain is that no service would survive that users did not value highly enough to pay for—and to pay for at prices that yielded the persons providing the service a more adequate income than alternative activities open to them. Neither the users nor the producers would be able to put their hands in anybody else's pocket to maintain a service that did not satisfy this condition.

FOOD AND DRUG ADMINISTRATION

By contrast with the ICC, the second major foray of the federal government into consumer protection—the Food and Drug Act of 1906—did not arise from protests over high prices, but from concern about the cleanliness of food. It was the era of the muckraker, of investigative journalism. Upton Sinclair had been sent by a socialist newspaper to Chicago to investigate conditions in the stockyards. The result was his famous novel, *The Jungle*, which he wrote to create sympathy for the workers, but which did far more to arouse indignation at the unsanitary conditions under which meat was processed. As Sinclair said at the time, "I aimed at the public's heart and by accident hit it in the stomach."

Long before *The Jungle* appeared and crystallized public sentiment in favor of legislation, such organizations as the Women's Christian Temperance Union and the National Temperance Society had formed the National Pure Food and Drug Congress (1898) to campaign for legislation to eliminate the medical nostrums of the day—mostly heavily laced with alcohol and so enabling spirits to be purchased and consumed in the guise of medicine, which explains the involvement of the temperance groups.

Here, too, special interests joined the reformers. The meat packers "learned very early in the history of the industry that it was not to their profit to poison their customers, especially in a competitive market in which the consumer could go elsewhere." They were especially concerned by restrictions on the importation of U.S. meat imposed by European countries, using as an excuse the allegation that the meat was diseased. They eagerly seized the opportunity to have the government certify that the meat was disease-free and at the same time pay for the inspection.[7]

Another special interest component was provided by the pharmacists and physicians through their professional associations, though their involvement was more complex and less single-mindedly economic than that of the meat packers—or of the railroads in the establishment of the ICC. Their economic interest was clear: patent medicines and nostrums, sold directly to the consumer by traveling medicine men and in other ways, competed with their services. Beyond this, they had a professional interest in the kinds of drugs and medicines available and were keenly aware of the dangers to the public from useless medicines promising miraculous cures for everything from cancer to leprosy. Public spirit and self-interest coincided.

The 1906 act was largely limited to the inspection of foods and the labeling of patent medicines, though, more by accident than design, it also subjected prescription drugs to control, a power which was not used until much later. The regulatory authority, from which the present Food and Drug Administration developed, was placed in the Department of Agriculture. Until the past fifteen years or so, neither the initial agency nor the FDA had much effect on the drug industry.

Few important new drugs were developed until sulfanilamide appeared in mid-1937. That was followed by the Elixir Sulfanilamide disaster, which occurred as a result of a chemist's efforts to make sulfanilamide available to patients who were unable to take capsules. The combination of the solvent he used and sulfanilamide proved deadly. By the end of the tragedy "a hundred and eight people were dead—a hundred and seven patients, who had taken the 'elixir,' and the chemist who had killed himself." [8] "Manufacturers themselves learned from the . . . experience the liability losses that could be suffered from the marketing of such drugs and instituted premarketing safety tests to avoid a repetition." [9] They also realized that government protection might be valuable to them. The result was the Food, Drug, and Cosmetic Act of 1938, which extended the government's control over advertising and labeling and required all new drugs to be approved for safety by the FDA before they could be sold in interstate commerce. Approval had to be granted or withheld within 180 days.

A cozy symbiotic relation developed between the pharmaceutical industry and the FDA until another tragedy occurred, the thalidomide episode of 1961–62. Thalidomide had been kept off the U.S. market by the FDA under the provisions of the 1938 act, though limited amounts of the drug have been distributed by physicians for experimental purposes. This limited distribution ended when reports surfaced about deformed babies born to European mothers who had taken thalidomide during pregnancy. The subsequent uproar swept into law in 1962 amendments that had developed out of Senator Kefauver's investigations of the drug industry the prior year. The tragedy also changed radically the thrust of the amendments. Kefauver had been concerned primarily with charges that drugs of dubious value were being sold at unduly high prices—the standard complaint about consumer exploitation by monopolistic business. As enacted, the amendments dealt more with quality than price. They "added a proof-of-efficacy requirement to the proof-of-safety requirement of the 1938 law, and they removed the time constraint on the F.D.A.'s disposition of a New Drug Application. No new drug may now be marketed unless and until the F.D.A. determines that there is substantial evidence not only that the drug is safe, as required under the 1938 law, but that it is effective in its intended use." [10]

The 1962 amendments coincided with the series of events that produced an explosion in government intervention and a change in its direction: the thalidomide tragedy, Rachel Carson's *Silent Spring*, which launched the environmental movement, and the controversy about Ralph Nader's *Unsafe at Any Speed*. The FDA participated in the changed role of government and became far more activist than it had ever been before. The banning of cyclamates and the threat to ban saccharin have received most public attention, but they are by no means the most important actions of the FDA.

No one can disagree with the objectives of the legislation that culminated in the 1962 amendments. Of course it is desirable that the public be protected from unsafe and useless drugs. However, it is also desirable that new drug development should be stimulated, and that new drugs should be made available to those who can benefit from them as soon as possible. As is so often the case, one good objective conflicts with other good objectives. Safety and caution in one direction can mean death in another.

The crucial questions are whether FDA regulation has been effective in reconciling these objectives and whether there may not be better ways of doing so. These questions have been studied in great detail. By now, considerable evidence has accumulated that indicates that FDA regulation is counterproductive, that it has done more harm by retarding progress in the production and distribution of valuable drugs than it has done good by preventing the distribution of harmful or ineffective drugs.

The effect on the rate of innovation of new drugs is dramatic: the number of "new chemical entities" introduced each year has fallen by more than 50 percent since 1962. Equally important, it now takes much longer for a new drug to be approved and, partly as a result, the cost of developing a new drug has been multiplied manyfold. According to one estimate for the 1950s and early 1960s, it then cost about half a million dollars and took about twenty-five months to develop a new drug and bring it to market. Allowing for inflation since then would raise the cost to a little over $1 million. By 1978, "it [was] costing $54 million and about eight years of effort to bring a drug to market"—a hundredfold increase in cost and quadrupling of time, compared with a doubling of prices in general.[11] As a result, drug companies can no longer afford to develop new drugs in the United States for patients with rare diseases. Increasingly, they must rely on drugs with high volume sales. The United States, long a leader in the development of new drugs, is rapidly taking a back seat. And we cannot even benefit fully from developments abroad because the FDA typically does not accept evidence from abroad as proof of effectiveness. The ultimate outcome may well be the same as in passenger rail traffic, the nationalization of the development of new drugs.

The so-called "drug lag" that has resulted is manifested in the relative availability of drugs in the United States and other countries. A careful study by Dr. William Wardell of the Center for the Study of Drug Development of the University of Rochester demonstrates, for example, that many more drugs are available in Great Britain that are not available in the United States than conversely, and that those available in both countries were on the average on the market sooner in Great Britain. Said Dr. Wardell in 1978,

> If you examine the therapeutic significance of drugs that haven't arrived in the U.S. but are available somewhere in the rest of the

world, such as in Britain, you can come across numerous examples where the patient has suffered. For example, there are one or two drugs called Beta blockers, which it now appears can prevent death after a heart attack—we call this secondary prevention of coronary death after myo-cardial infarction—which, if available here, could be saving about ten thousand lives a year in the United States. In the ten years after the 1962 amendments, no drug was approved for hypertension—that's for the control of blood pressure—in the United States, whereas several were approved in Britain. In the entire cardiovascular area, only one drug was approved in the five year period from '67 to '72. And this can be correlated with known organizational problems at F.D.A. . . .

The implications for the patient are that therapeutic decisions that used to be the preserve of the doctor and the patient are increasingly being made at a national level, by committees of experts, and these committees and the agency for which they are acting—the F.D.A.— are highly skewed towards avoiding risks so there's a tendency for us to have drugs that are safer but not to have drugs that are effective. Now I've heard some remarkable statements from some of these advisory committees where in considering drugs one has seen the statement "there are not enough patients with a disease of this severity to warrant marketing this drug for general use." Now that's fine if what you are trying to do is minimize drug toxicity for the whole population, but if you happen to be one of those "not enough patients," and you have a disease that is of high severity or a disease that's very rare, then that's just tough luck for you.

Granted all this, may these costs not be justified by the advantage of keeping dangerous drugs off the market, of preventing a series of thalidomide disasters? The most careful empirical study of this question that has been made, by Sam Peltzman, concludes that the evidence is unambiguous: that the harm done has greatly outweighed the good. He explains his conclusion partly by noting that "the penalties imposed by the marketplace on sellers of ineffective drugs before 1962 seems to have been sufficient to have left little room for improvement by a regulatory agency." [12] After all, the manufacturers of thalidomide ended up paying many tens of millions of dollars in damages—surely a strong incentive to avoid any similar episodes. Of course, mistakes will still happen— the thalidomide tragedy was one—but so will they under government regulation.

The evidence confirms what general reasoning strongly suggests. It is no accident that the FDA, despite the best of intentions, operates to discourage the development and prevent the marketing of new and potentially useful drugs.

Put yourself in the position of an FDA official charged with approving or disapproving a new drug. You can make two very different mistakes:

1. Approve a drug that turns out to have unanticipated side effects resulting in the death or serious impairment of a sizable number of persons.

2. Refuse approval of a drug that is capable of saving many lives or relieving great distress and that has no untoward side effects.

If you make the first mistake—approve a thalidomide—your name will be spread over the front page of every newspaper. You will be in deep disgrace. If you make the second mistake, who will know it? The pharmaceutical firm promoting the new drug, which will be dismissed as an example of greedy businessmen with hearts of stone, and a few disgruntled chemists and physicians involved in developing and testing the new product. The people whose lives might have been saved will not be around to protest. Their families will have no way of knowing that their loved ones lost their lives because of the "caution" of an unknown FDA official.

In view of the contrast between the abuse poured on the European drug companies that sold thalidomide and the fame and acclaim that came to the woman who held up approval of thalidomide in the United States (Dr. Frances O. Kelsey, given a gold medal for Distinguished Government Service by John F. Kennedy), is there any doubt which mistake you will be more anxious to avoid? With the best will in the world, you or I, if we were in that position, would be led to reject or postpone approval of many a good drug in order to avoid even a remote possibility of approving a drug that will have newsworthy side effects.

This inevitable bias is reinforced by the reaction of the pharmaceutical industry. The bias leads to unduly stringent standards. Getting approval becomes more expensive, time-consuming, and risky. Research on new drugs becomes less profitable. Each company has less to fear from the research efforts of its competitors.

Existing firms and existing drugs are protected from competition. New entry is discouraged. Research that is done will be concentrated on the least controversial, which means least innovative, of the new possibilities.

When one of us suggested in a *Newsweek* column (January 8, 1973) that for these reasons the FDA should be abolished, the column evoked letters from persons in pharmaceutical work offering tales of woe to confirm the allegation that the FDA was frustrating drug development. But most also said something like, "In contrast to your opinion, I do not believe that the FDA should be abolished but I do believe that its power should be" changed in such and such a way.

A subsequent column, entitled "Barking Cats" (February 19, 1973), replied:

> What would you think of someone who said, "I would like to have a cat provided it barked"? Yet your statement that you favor an FDA provided it behaves as you believe desirable is precisely equivalent. The biological laws that specify the characteristics of cats are no more rigid than the political laws that specify the behavior of governmental agencies once they are established. The way the FDA now behaves, and the adverse consequences, are not an accident, not a result of some easily corrected human mistake, but a consequence of its constitution in precisely the same way that a meow is related to the constitution of a cat. As a natural scientist, you recognize that you cannot assign characteristics at will to chemical and biological entities, cannot demand that cats bark or water burn. Why do you suppose the situation is different in the social sciences?

The error of supposing that the behavior of social organisms can be shaped at will is widespread. It is the fundamental error of most so-called reformers. It explains why they so often feel that the fault lies in the man, not the "system"; that the way to solve problems is to "turn the rascals out" and put well-meaning people in charge. It explains why their reforms, when ostensibly achieved, so often go astray.

The harm done by the FDA does not result from defects in the people in charge—unless it be a defect to be human. Many have been able and devoted civil servants. However, social, political, and economic pressures determine the behavior of the people supposedly in charge of a government agency to a far greater extent than they determine its behavior. No doubt there are exceptions, but they are rare—almost as rare as barking cats. That does not mean that effective reform is impossible. But it requires taking account of the political laws governing the behavior of government agencies, not simply berating officials for inefficiency and waste or questioning their motives and urging them to do better. The FDA did far less harm than it does now before the Kefauver amendments altered the pressures and incentives of the civil servants.

CONSUMER PRODUCTS SAFETY COMMISSION

The Consumer Products Safety Commission exemplifies the change in regulatory activity in the past decade or so. It cuts across industries. Its main concern is not with price or cost but with safety. It has wide discretionary authority and operates under only the most general of mandates.

Activated on May 14, 1973, "[t]he Commission is specifically mandated to protect the public against unreasonable risks of injury from consumer products, to assist consumers in evaluating the safety of these products, to develop standards for consumer products, to minimize conflicts of these standards at the Federal, state and local level, and to promote research and investigation into the causes and prevention of product-related deaths, illnesses, and injuries." [13]

Its authority covers "any article or component part produced or distributed (i) for sale to a consumer . . . or (ii) for the personal use, consumption or enjoyment of a consumer" except for "tobacco and tobacco products; motor vehicles and motor vehicle equipment; drugs; food; aircraft and aircraft components; certain boats; and certain other items"—almost all covered by such other regulatory agencies as the Bureau of Alcohol, Tobacco and Firearms, the National Highway Traffic Safety Administration, the FDA, the Federal Aviation Administration, and the Coast Guard. [14]

Although the CPSC is in its early stages, it is likely to become a major agency that will have far-reaching effects on the products and services we shall be able to buy. It has conducted tests and issued standards on products varying from book matches to bicycles, from children's toy cap guns to television receivers, from refuse bins to miniature Christmas tree lights.

The objective of safer products is obviously a good one, but at what cost and by what standards? "Unreasonable risk" is hardly a scientific term capable of objective specification. What decibel level of noise from a cap gun is an "unreasonable risk" to a child's (or adult's) hearing? The spectacle of trained, highly paid "experts" with ear muffs shooting cap guns as part of the process of trying to answer that question is hardly calculated to instill confidence in the taxpayer that his money is being spent sensibly. A "safer" bicycle may be slower, heavier, and costlier than a less "safe" bicycle. By what criteria can the CPSC bureaucrats, in issuing their standards, decide how much speed to sacrifice, how much weight to add, how much extra cost to impose in order to achieve how much extra safety? Do "safer" standards produce more safety? Or do they only encourage less attention and care by the user? Most bicycle and similar accidents are, after all, caused by human carelessness or error.

Most of these questions do not admit of objective answers—yet they must be answered implicitly in the course of devising and issuing standards. The answers will reflect partly the arbitrary judgments of the civil servants involved, occasionally the judgment of consumers or consumer organizations that happen to have a special interest in the item in question, but mostly the influence of the makers of the products. In the main, they are the only ones who have sufficient interest and expertise to comment knowledgeably on proposed standards. Indeed, much of the formulation of standards has simply been turned over to trade associations. You may be sure those standards will be formulated in the interest of the members of the association, with a sharp eye to protecting themselves from competition, both from possible new producers at home and from foreign producers. The result will be to strengthen the competitive position of existing domestic manufacturers and to make innovation and the development of new and improved products more expensive and difficult.

When products enter the marketplace in the usual course of events, there is an opportunity for experiment, for trial and error.

No doubt, shoddy products are produced, mistakes are made, un-suspected defects turn up. But mistakes usually tend to be on a small scale—though some are major, as in the recent case of the Firestone 500 radial tire—and can be corrected gradually. Consumers can experiment for themselves, decide what features they like and what features they do not like.

When the government steps in through CPSC, the situation is different. Many decisions must be made before the product has been subjected to extensive trial and error in actual use. The standards cannot be adjusted to different needs and tastes. They must apply uniformly to all. Consumers will inevitably be denied the opportunity to experiment with a range of alternatives. Mistakes will still be made, and when they are, they are almost sure to be major.

Two examples from the CPSC illustrate the problem.

In August 1973, only three months after starting operation, it "banned certain brands of aerosol spray adhesives as an imminent hazard. Its decision was based primarily on the preliminary findings of one academic researcher who claimed that they could cause birth defects. After more thorough research failed to corroborate the initial report, the commission lifted the ban in March 1974." [15]

That prompt admission of error is most commendable and most unusual for a government agency. Yet it did not prevent harm. "It seems that at least nine pregnant women who had used the spray adhesives reacted to the news of the commission's initial decision by undergoing abortions. They decided not to carry through their pregnancies for fear of producing babies with birth defects." [16]

A far more serious example is the episode with respect to Tris. The commission, when established, was assigned responsibility for administering the "Flammable Fabrics Act," dating back to 1953, which was intended to reduce death and injuries from the accidental burning of products, fabrics, or related materials. A standard for children's sleepwear that had been issued in 1971 by the predecessor agency was strengthened by the CPSC in mid-1973. At the time the cheapest way to meet this standard was by impregnating the cloth with a flame-retardant chemical—

Tris. Soon, something like 99 percent of all children's sleepwear produced and sold in the United States was impregnated with Tris. Later it was discovered that Tris was a potent carcinogen. On April 8, 1977, the commission banned its use in children's apparel and provided for withdrawal of Tris-treated garments from the market and their return by consumers.

Needless to say, in its 1977 *Annual Report* the commission made a virtue of the correction of a dangerous situation that had arisen solely as a result of its own earlier actions, without acknowledging its own role in the development of the problem. The initial requirements exposed millions of children to the danger of developing cancer. Both the initial requirements and the subsequent banning of Tris imposed heavy costs on the producers of children's sleepwear, which meant, ultimately, on their customers. They were taxed, as it were, coming and going.

This example is instructive in showing the difference between across-the-board regulation and the operation of the market. Had the market been allowed to operate, some manufacturers no doubt would have used Tris in order to try to enhance the appeal of their sleepwear by being able to claim flame resistance, but Tris would have been introduced gradually. There would have been time for the information about Tris's carcinogenic qualities to have been discovered and to lead to its withdrawal before it was used on a massive scale.

ENVIRONMENT

The environmental movement is responsible for one of the most rapidly growing areas of federal intervention. The Environmental Protection Agency, established in 1970 "to protect and enhance the physical environment," has been granted increasing power and authority. Its budget has multiplied sevenfold from 1970 to 1978 and is now more than half a billion dollars. It has a staff of about 7,000.[17] It has imposed costs on industry and local and state governments to meet its standards that total in the tens of billions of dollars a year. Something between a tenth and a quarter of total net investment in new capital equipment by business now goes for antipollution purposes. And this does not count the costs of requirements imposed by other agencies, such as those designed to control emissions of motor vehicles, or the costs of land-use planning or wilderness preservation or a host of other federal, state, and local government activities undertaken in the name of protecting the environment.

The preservation of the environment and the avoidance of undue pollution are real problems and they are problems concerning which the government has an important role to play. When all the costs and benefits of any action, and the people hurt or benefited, are readily identifiable, the market provides an excellent means for assuring that only those actions are undertaken for which the benefits exceed the costs for all participants. But when the costs and benefits or the people affected cannot be identified, there is a market failure of the kind discussed in Chapter 1 as arising from "third-party" or neighborhood effects.

To take a simple example, if someone upstream contaminates a river, he is, in effect, exchanging bad water for good water with people downstream. There may well be terms on which the people downstream would be willing to make the exchange. The problem is that it isn't feasible to make that transaction the subject of a voluntary exchange, to identify just who got the bad water that a particular person upstream was responsible for, and to require that his permission be obtained.

Government is one means through which we can try to compensate for "market failure," try to use our resources more effectively to produce the amount of clean air, water, and land that we are willing to pay for. Unfortunately, the very factors that produce the market failure also make it difficult for government to achieve a satisfactory solution. Generally, it is no easier for government to identify the specific persons who are hurt and benefited than for market participants, no easier for government to assess the amount of harm or benefit to each. Attempts to use government to correct market failure have often simply substituted government failure for market failure.

Public discussion of the environmental issue is frequently characterized more by emotion than reason. Much of it proceeds as if the issue is pollution versus no pollution, as if it were desirable and possible to have a world without pollution. That is clearly nonsense. No one who contemplates the problem seriously will regard zero pollution as either a desirable or a possible state of affairs. We could have zero pollution from automobiles, for example, by simply abolishing all automobiles. That would also make the kind of agricultural and industrial productivity we now enjoy impossible, and so condemn most of us to a drastically lower standard of living, perhaps many even to death. One source of atmospheric pollution is the carbon dioxide that we all exhale. We could stop that very simply. But the cost would clearly exceed the gain.

It costs something to have clean air, just as it costs something to have other good things we want. Our resources are limited and we must weigh the gains from reducing pollution against the costs. Moreover, "pollution" is not an objective phenomenon. One person's pollution may be another's pleasure. To some of us rock music is noise pollution; to others of us it is pleasure.

The real problem is not "eliminating pollution," but trying to establish arrangements that will yield the "right" amount of pollution: an amount such that the gain from reducing pollution a bit more just balances the sacrifice of the other good things— houses, shoes, coats, and so on—that would have to be given up in order to reduce the pollution. If we go farther than that, we sacrifice more than we gain.

Another obstacle to rational analysis of the environmental issue is the tendency to pose it in terms of good or evil—to proceed as if bad, malicious people are pouring pollutants into the atmosphere out of the blackness of their hearts, that the problem is one of motives, that if only those of us who are noble would rise in our wrath to subdue the evil men, all would be well. It is always much easier to call other people names than to engage in hard intellectual analysis.

In the case of pollution, the devil blamed is typically "business," the enterprises that produce goods and services. In fact, the people responsible for pollution are consumers, not producers. They create, as it were, a demand for pollution. People who use electricity are responsible for the smoke that comes out of the stacks of the generating plants. If we want to have the electricity with less pollution, we shall have to pay, directly or indirectly, a high enough price for the electricity to cover the extra costs. Ultimately, the cost of getting cleaner air, water, and all the rest must be borne by the consumer. There is no one else to pay for it. Business is only an intermediary, a way of coordinating the activities of people as consumers and producers.

The problem of controlling pollution and protecting the environment is greatly complicated by the tendency for the gains and losses derived from doing so to fall on different people. The people, for example, who gain from the greater availability of wilderness areas, or from the improvement of the recreational quality of lakes or rivers, or from the cleaner air in the cities, are generally not the same people as those who would lose from the resulting higher costs of food or steel or chemicals. Typically, we suspect, the people who would benefit most from the reduction of pollution are better off, financially and educationally, than the people who would benefit most from the lower cost of things that would result from permitting more pollution. The latter might prefer cheaper electricity to cleaner air. Director's Law is not absent from the pollution area.

The same approach has generally been adopted in the attempt to control pollution as in regulating railroads and trucks, controlling food and drugs, and promoting the safety of products. Establish a government regulatory agency that has discretionary power to issue rules and orders specifying actions that private enterprises or individuals or states and local communities must take. Seek to enforce these regulations by sanctions imposed by the agency or by courts.

This system provides no effective mechanism to assure the balancing of costs and benefits. By putting the whole issue in terms of enforceable orders, it creates a situation suggestive of crime and punishment, not of buying and selling; of right and wrong, not of more or less. Moreover, it has the same defects as this kind of regulation in other areas. The persons or agencies regulated have a strong interest in spending resources, not to achieve the desired objectives, but to get favorable rulings, to influence the bureaucrats. And the self-interest of the regulators in its turn bears only the most distant relation to the basic objective. As always in the bureaucratic process, diffused and widely spread interests get short shrift; the concentrated interests take over. In the past these have generally been the business enterprises, and particularly the large and important ones. More recently they have been joined by the self-styled, highly organized "public interest" groups that profess to speak for a constituency that may be utterly unaware of their existence.

Most economists agree that a far better way to control pollution than the present method of specific regulation and supervision is to introduce market discipline by imposing effluent charges. For example, instead of requiring firms to erect specific kinds of waste disposal plants or to achieve a specified level of water quality in water discharged into a lake or river, impose a tax of a specified amount per unit of effluent discharged. That way, the firm would have an incentive to use the cheapest way to keep down the effluent. Equally important, that way there would be objective evidence of the costs of reducing pollution. If a small tax led to a large reduction, that would be a clear indication that there is little to gain from permitting the discharge. On the other hand, if even a high tax left much discharge, that would indicate the reverse, but also would provide substantial sums to compensate the losers or undo the damage. The tax rate itself could be varied as experience yielded information on costs and gains.

Like regulations, an effluent charge automatically puts the cost on the users of the products responsible for the pollution. Those products for which it is expensive to reduce pollution would go up in price compared to those for which it is cheap, just as now those products on which regulations impose heavy costs go up in price relative to others. The output of the former would go down, of the latter up. The difference between the effluent charge and the regulations is that the effluent charge would control pollution more effectively at lower cost, and impose fewer burdens on nonpolluting activities.

In an excellent article A. Myrick Freeman III and Robert H. Haveman write, "It is not entirely facetious to suggest that the reason an economic-incentive approach has not been tried in this country is that it would work."

As they say, "Establishment of a pollution-charge system in conjunction with environmental quality standards would resolve most of the political conflict over the environment. And it would do so in a highly visible way, so that those who would be hurt by such a policy could see what was happening. It is the openness and explicitness of such choices that policy makers seek to avoid." [18]

This is a very brief treatment of an extremely important and far-reaching problem. But perhaps it is sufficient to suggest that the difficulties that have plagued government regulation in areas where government has no place whatsoever—as in fixing prices and allocating routes in trucking, rail travel, and air travel—also arise in areas where government has a role to play.

Perhaps also it may lead to a second look at the performance of market mechanisms in areas where they admittedly operate imperfectly. The imperfect market may, after all, do as well or better than the imperfect government. In pollution, such a look would bring many surprises.

If we look not at rhetoric but at reality, the air is in general far cleaner and the water safer today than one hundred years ago. The air is cleaner and the water safer in the advanced countries of the world today than in the backward countries. Industrialization has raised new problems, but it has also provided the means to solve prior problems. The development of the automobile did add to one form of pollution—but it largely ended a far less attractive form.

DEPARTMENT OF ENERGY

The embargo of the United States instituted by the OPEC cartel in 1973 ushered in a series of energy crises and occasional long lines at gasoline stations that have plagued us ever since. Government has reacted by establishing one bureaucratic organization after another to control and regulate energy production and use, terminating in the establishment of a Department of Energy in 1977.

Government officials, newspaper reports, and TV commentators regularly attribute the energy crisis to a rapacious oil industry, or wasteful consumers, or bad weather, or Arab sheikhs. But none of these is responsible.

After all, the oil industry has been around for a long time—

and has always been rapacious. Consumers have not suddenly become wasteful. We have had hard winters before. Arab sheikhs have desired wealth as far back as human memory runs.

The subtle and sophisticated people who fill the newspaper columns and the airwaves with such silly explanations seem never to have asked themselves the obvious question: why is it that for a century and more before 1971, there were no energy crises, no gasoline shortages, no problems about fuel oil—except during World War II?

There has been an energy crisis because government created one. Of course, government has not done so deliberately. Presidents Nixon, Ford, or Carter never sent a message to Congress asking it to legislate an energy crisis and long gasoline lines. But he who says A must say B. Ever since President Nixon froze wages and prices on August 15, 1971, the government has imposed maximum prices on crude oil, gasoline at retail, and other petroleum products. Unfortunately, the quadrupling of crude oil prices by the OPEC cartel in 1973 prevented those maximum prices from being abolished when all others were. Maximum legal prices for petroleum products—that is the key element common both to World War II and the period since 1971.

Economists may not know much. But we know one thing very well: how to produce surpluses and shortages. Do you want a surplus? Have the government legislate a *minimum* price that is *above* the price that would otherwise prevail. That is what we have done at one time or another to produce surpluses of wheat, of sugar, of butter, of many other commodities.

Do you want a shortage? Have the government legislate a *maximum* price that is *below* the price that would otherwise prevail. That is what New York City and, more recently, other cities have done for rental dwellings, and that is why they all suffer or will soon suffer from housing shortages. That is why there were so many shortages during World War II. That is why there is an energy crisis and a gasoline shortage.

There is one simple way to end the energy crisis and gasoline shortages tomorrow—and we mean tomorrow and not six months from now, not six years from now. Eliminate all controls on the prices of crude oil and other petroleum products.

Other misguided policies of government and the monopolistic

behavior of the OPEC cartel might keep petroleum products expensive, but they would not produce the disorganization, chaos, and confusion that we now confront.

Perhaps surprisingly, this solution would reduce the cost of gasoline to the consumer—the *true* cost. Prices at the pump might go up a few cents a gallon, but the cost of gasoline includes the time and gasoline wasted standing in line and hunting for a gas station with gas, *plus* the annual budget of the Department of Energy, which amounted to $10.8 billion in 1979, or to around 9 cents per gallon of gasoline.

Why has this simple and foolproof solution not been adopted? So far as we can see, for two basic reasons—one general, the other specific. To the despair of every economist, it seems almost impossible for most people other than trained economists to comprehend how a price system works. Reporters and TV commentators seem especially resistant to the elementary principles they supposedly imbibed in freshman economics. Second, removing price controls would reveal that the emperor is naked—it would show how useless, indeed harmful, are the activities of the 20,000 employees of the Department of Energy. It might even occur to someone how much better off we were before we had a Department of Energy.

But what about the claim by President Carter that the government must institute a massive program to produce synthetic fuels or else the nation will run out of energy by 1990? That, too, is a myth. A government program seems the solution only because government has been blocking at every turn the effective free market solution.

We pay OPEC nations around $20 a barrel for oil under long-term contracts and even more on the spot market (the market for immediate delivery), but the government forces domestic producers to sell oil for as little as $5.94 a barrel. Government taxes the domestic production of oil to subsidize oil imported from abroad. We pay more than twice as much for imported liquefied natural gas from Algeria as the government permits domestic producers of natural gas to charge. Government imposes stringent environmental requirements on both users and producers of energy with little or no regard to the economic costs involved. Complicated rules and red tape add greatly to the time required to build power plants whether nuclear, oil, or coal, and to bring into production our abundant supply of coal—and multiply the cost. These counterproductive government policies have stifled domestic production of energy and have made us more dependent than ever on foreign oil—despite, as President Carter put it, "the danger of depending on a thin line of oil tankers stretching halfway around the world."

In mid-1979, President Carter proposed a massive government program stretching over a decade and costing $88 billion to produce synthetic fuel. Does it really make sense to commit the taxpayers to spending, directly or indirectly, $40 or more for a barrel of oil from shale while prohibiting the owners of domestic wells from receiving more than $5.94 for some categories of oil? Or, as Edward J. Mitchell put it in a *Wall Street Journal* article (August 27, 1979), "We may well question . . . how spending $88 billion to obtain a modest amount of $40 per barrel synthetic oil in 1990 'protects' us from $20 per barrel OPEC oil either today or in 1990."

Fuel from shale, tar sands, and so on, makes sense if and only if that way to produce energy is cheaper than alternatives—account being taken of all costs. The most effective mechanism to determine whether it is cheaper is the market. If it is cheaper, it will be in the self-interest of private enterprises to exploit these alternatives—provided they reap the benefits and bear the cost.

Private enterprises can count on reaping the benefits only if they are confident that future prices will not be controlled. Otherwise, they are asked to engage in a heads-you-win, tails-I-lose gamble. That is the present situation. If the price rises, controls and "windfall taxes" loom; if the price falls, they hold the bag. That prospect emasculates the free market and makes President Carter's socialist policy the only alternative.

Private enterprises will bear all the cost only if they are required to pay for environmental damage. The way to do that is to impose effluent charges—not to have one government agency impose arbitrary standards and then set up another to cut through the first's red tape.

The threat of price control and regulation is the only important obstacle to the development of alternative fuels by private enterprise. It is argued that the risks are too great and the capital costs too heavy. That is simply wrong. Risk taking is the essence of private enterprise. Risks are not eliminated by imposing them on the taxpayer instead of on the capitalist. And the Alaska pipeline shows that private markets can raise massive sums for promising projects. The capital resources of the nation are not increased by using the tax collector rather than the stock market to mobilize them.

The bottom line is that come what may, we the people shall pay for the energy we consume. And we shall pay far less in total, and have far more energy, if we pay directly and are left free to choose for ourselves how to use energy than if we pay indirectly through taxes and inflation and are told by government bureaucrats how to use energy.

THE MARKET

Perfection is not of this world. There will always be shoddy products, quacks, con artists. But on the whole, market competition, when it is permitted to work, protects the consumer better than do the alternative government mechanisms that have been increasingly superimposed on the market.

As Adam Smith said in the quotation with which we began this chapter, competition does not protect the consumer because businessmen are more soft-hearted than the bureaucrats or because they are more altruistic or generous, or even because they are more competent, but only because it is in the self-interest of the businessman to serve the consumer.

If one storekeeper offers you goods of lower quality or of higher price than another, you're not going to continue to patronize his store. If he buys goods to sell that don't serve your needs, you're not going to buy them. The merchants therefore search out all over the world the products that might meet your needs and might appeal to you. And they stand back of them because if they don't, they're going to go out of business. When you enter a store, no one forces you to buy. You are free to do so or go elsewhere. That is the basic difference between the market and a political agency. You are free to choose. There is no policeman to take the money out of your pocket to pay for something you do not want or to make you do something you do not want to do.

But, the advocate of government regulation will say, suppose the FDA weren't there, what would prevent business from distributing adulterated or dangerous products? It would be a very expensive thing to do—as the examples of Elixir Sulfanilamide and thalidomide and numerous less publicized incidents indicate. It is very poor business practice—not a way to develop a loyal and faithful clientele. Of course, mistakes and accidents occur—but as the Tris case illustrates, government regulation doesn't prevent them. The difference is that a private firm that makes a serious blunder may go out of business. A government agency is likely to get a bigger budget.

Cases will arise where adverse effects develop that could not have been foreseen—but government has no better means of predicting such developments than private enterprise. The only way to prevent all such developments would be to stop progress, which would also eliminate the possibility of unforeseen favorable developments.

But, the advocate of government regulation will say, without the Consumer Products Safety Commission, how can the consumer judge the quality of complex products? The market's answer is that he does not have to be able to judge for himself. He has other bases for choosing. One is the use of a middleman. The chief economic function of a department store, for example, is to monitor quality on our behalf. None of us is an expert on all of the items we buy, even the most trivial, like shirts, ties, or shoes. If we buy an item that turns out to be defective, we are more likely to return it to the retailer from whom we bought it than to the manufacturer. The retailer is in a far better position to judge quality than we are. Sears, Roebuck and Montgomery Ward, like department stores, are effective consumer testing and certifying agencies as well as distributors.

Another market device is the brand name. It is in the self-interest of General Electric or General Motors or Westinghouse or Rolls-Royce to get a reputation for producing dependable, reliable products. That is the source of their "goodwill," which may well contribute more to their value as a firm than the factories and plants they own.

Still another device is the private testing organization. Such testing laboratories are common in industry and serve an extremely important role in certifying the quality of a vast array of products. For the consumer there are private organizations like Consumers' Research, started in 1928, and still in business reporting evaluations of a wide range of consumer products in its monthly *Consumers' Research* magazine; and Consumers Union, founded in 1935, which publishes *Consumer Reports*.

Both Consumers' Research and Consumers Union have been highly successful—enough so to maintain sizable staffs of engineers and other trained testing and clerical personnel. Yet after nearly half a century, they have been able to attract at most 1 or 2 percent of the potential clientele. Consumers Union, the larger of the two, has about 2 million members. Their existence is a market response to consumer demand. Their small size and the failure of other such agencies to spring up demonstrates that only a small minority of consumers demand and are willing to pay for such a service. It must be that most consumers are getting the guidance they want and are willing to pay for in some other way.

What about the claim that consumers can be led by the nose by advertising? Our answer is that they can't—as numerous expensive advertising fiascos testify. One of the greatest duds of all time was the Edsel automobile, introduced by Ford Motor Company and promoted by a major advertising campaign. More basically, advertising is a cost of doing business, and the businessman wants to get the most for his money. Is it not more sensible to try to appeal to the real wants or desires of consumers than to try to manufacture artificial wants or desires? Surely it will generally be cheaper to sell them something that meets wants they already have than to create an artificial want.

A favorite example has been the allegedly artificially created desire for automobile model changes. Yet Ford was unable to make a success of the Edsel despite an enormously expensive advertising campaign. There always have been cars available that did not make frequent model changes—the Superba in the United States (the passenger counterpart of the Checker cab), and many foreign cars. They were never able to attract more than a small fraction of the total custom. If that was what consumers *really* wanted, the companies that offered that option would have prospered, and the others would have followed suit. The real objection of most critics of advertising is not that advertising manipulates tastes but that the public at large has meretricious tastes—that is, tastes that do not agree with the critics'.

In any event, you cannot beat something with nothing. One must always compare alternatives: the real with the real. If business advertising is misleading, is no advertising, or government control of advertising, preferable? At least with private business there is competition. One advertiser can dispute another. That is more difficult with government. Government, too, engages in advertising. It has thousands of public relations agents to present its product in the most favorable light. That advertising is often more misleading than anything put out by private enterprises. Consider only the advertising the Treasury uses to sell its savings bonds: "United States Savings Bonds . . . What a great way to save!" as the slogan goes on a slip produced by the U.S. Treasury Department and distributed by banks to their customers. Yet anyone who has bought government savings bonds over the past decade and more has been taken to the cleaners. The amount he received on maturity would buy less in goods and services than the amount he paid for the bond, and he has had to pay taxes on the mislabeled "interest." And all this because of inflation produced by the government that sold him the bonds! Yet the Treasury continues to advertise the bonds as "building personal security," as a "gift that keeps on growing," to quote further from the same slip.

What about the danger of monopoly that led to the antitrust laws? That is a real danger. The most effective way to counter it is not through a bigger antitrust division at the Department of Justice or a larger budget for the Federal Trade Commission, but through removing existing barriers to international trade. That would permit competition from all over the world to be even more effective than it is now in undermining monopoly at home. Freddie Laker of Britain needed no help from the Department of Justice to crack the airline cartel. Japanese and German automobile manufacturers forced American manufacturers to introduce smaller cars.

The great danger to the consumer is monopoly—whether private or governmental. His most effective protection is free competition at home and free trade throughout the world. The consumer is protected from being exploited by one seller by the existence of another seller from whom he can buy and who is eager to sell to him. Alternative sources of supply protect the consumer far more effectively than all the Ralph Naders of the world.

CONCLUSION

"The reign of tears is over. The slums will be only a memory. We will turn our prisons into factories and our jails into storehouses and corncribs. Men will walk upright now, women will smile, and the children will laugh. Hell will be forever for rent." [19]

That is how Billy Sunday, noted evangelist and leading crusader against Demon Rum, greeted the onset of Prohibition in 1920, enacted in a burst of moral righteousness at the end of the First World War. That episode is a stark reminder of where the present burst of moral righteousness, the present drive to protect us from ourselves, can lead.

Prohibition was imposed for our own good. Alcohol *is* a dangerous substance. More lives are lost each year from alcohol than from all the dangerous substances the FDA controls put together. But where did Prohibition lead?

New prisons and jails had to be built to house the criminals spawned by converting the drinking of spirits into a crime against the state. Al Capone, Bugs Moran became notorious for their exploits—murder, extortion, hijacking, bootlegging. Who were their customers? Who bought the liquor they purveyed illegally? Respectable citizens who would never themselves have approved of, or engaged in, the activities that Al Capone and his fellow gangsters made infamous. They simply wanted a drink. In order to have a drink, they had to break the law. Prohibition didn't stop drinking. It did convert a lot of otherwise law-obedient citizens into lawbreakers. It did confer an aura of glamour and excitement to drinking that attracted many young persons. It did suppress many of the disciplinary forces of the market that ordinarily protect the consumer from shoddy, adulterated, and dangerous products. It did corrupt the minions of the law and create a decadent moral climate. It did *not* stop the consumption of alcohol.

We are as yet a long way from that today, with the prohibition of cyclamates, DDT, and laetrile. But that is the direction in which we are headed. Something of a gray market already exists in drugs that are prohibited by the FDA; citizens already go to Canada or Mexico to buy drugs they cannot legally buy in the United States—just as people did during Prohibition to get a legal drink. Many a conscientious physician feels himself in a dilemma, caught between what he regards as the welfare of his patient and strict obedience to the law.

If we continue on this path, there is no doubt where it will end. If the government has the responsibility of protecting us from dangerous substances, the logic surely calls for prohibiting alcohol and tobacco. If it is appropriate for the government to protect us from using dangerous bicycles and cap guns, the logic calls for prohibiting still more dangerous activities such as hang-gliding, motorcycling, and skiing.

Even the people who administer the regulatory agencies are appalled at this prospect and withdraw from it. As for the rest of us, the reaction of the public to the more extreme attempts to control our behavior—to the requirement of an interlock system on automobiles or the proposed ban of saccharin—is ample evidence that we want no part of it. Insofar as the government has information not generally available about the merits or demerits of the items we ingest or the activities we engage in, let it give us the information. But let it leave us free to choose what chances we want to take with our own lives.

CHAPTER 8

Who Protects
the Worker?

Over the past two centuries the condition of the ordinary worker in the United States and other economically advanced societies has improved enormously. Hardly any worker today engages in the kind of backbreaking labor that was common a century or so ago and that is still common over most of the globe. Working conditions are better; hours of work are shorter; vacations and other fringe benefits are taken for granted. Earnings are far higher, enabling the ordinary family to achieve a level of living that only the affluent few could earlier enjoy.

If Gallup were to conduct a poll asking: "What accounts for the improvement in the lot of the worker?" the most popular answer would very likely be "labor unions," and the next, "government"—though perhaps "no one" or "don't know" or "no opinion" would beat both. Yet the history of the United States and other Western countries over the past two centuries demonstrates that these answers are wrong.

During most of the period, unions were of little importance in the United States. As late as 1900, only 3 percent of all workers were members of unions. Even today fewer than one worker in four is a member of a union. Unions were clearly not a major reason for the improvement in the lot of the worker in the United States.

Similarly, until the New Deal, regulation of and intervention in economic arrangements by government, and especially central government, were minimal. Government played an essential role by providing a framework for a free market. But direct government action was clearly not the reason for the improvement in the lot of the worker.

As to "no one" accounting for the improvement, the very lot of the worker today belies that answer.

LABOR UNIONS

One of the most egregious misuses of language is the use of "labor" as if it were synonymous with "labor unions"—as in reports that "labor opposes" such and such a proposed law or that the legislative program of "labor" is such and such. That is a double error. In the first place, more than three out of four workers in the United States are not members of labor unions. Even in Great Britain, where labor unions have long been far stronger than in the United States, most workers are not members of labor unions. In the second place, it is an error to identify the interests of a "labor union" with the interests of its members. There is a connection, and a close connection, for most unions most of the time. However, there are enough cases of union officials acting to benefit themselves at the expense of their members, both in legal ways and by misuse and misappropriation of union funds, to warn against the automatic equating of the interests of "labor unions" with the interests of "labor union members," let alone with the interests of labor as a whole.

This misuse of language is both a cause and an effect of a general tendency to overestimate the influence and role of labor unions. Union actions are visible and newsworthy. They often generate front-page headlines and full-scale coverage on the nightly TV programs. "The higgling and bargaining of the market"—as Adam Smith termed it—whereby the wages of most workers in the United States are determined is far less visible, draws less attention, and its importance is as a result greatly underestimated.

The misuse of language contributes also to the belief that labor unions are a product of modern industrial development. They are nothing of the kind. On the contrary, they are a throwback to a preindustrial period, to the guilds that were the characteristic form of organization of both merchants and craftsmen in the cities and city-states that grew out of the feudal period. Indeed, the modern labor union can be traced back even further, nearly 2,500 years to an agreement reached among medical men in Greece.

Hippocrates, universally regarded as the father of modern medicine, was born around 460 B.C. on the island of Cos, one of the Greek islands only a few miles away from the coast of Asia Minor. At the time it was a thriving island, and already a medical center. After studying medicine on Cos, Hippocrates traveled widely, developing a great reputation as a physician, particularly for his ability to end plagues and epidemics. After a time he returned to Cos, where he established, or took charge of, a medical school and healing center. He taught all who wished to learn— so long as they paid the fees. His center became famous throughout the Greek world, attracting students, patients, and physicians from far and wide.

When Hippocrates died at the age of 104, or so legend has it, Cos was full of medical people, his students and disciples. Competition for patients was fierce and, not surprisingly, a concerted movement apparently developed to do something about it—in modern terminology, to "rationalize" the discipline in order to eliminate "unfair competition."

Accordingly, some twenty years or so after Hippocrates died— again, as legend has it—the medical people got together and constructed a code of conduct. They named it the Hippocratic Oath after their old teacher and master. Thereafter, on the island of Cos and increasingly throughout the rest of the world, every newly trained physician, before he could start practice, was required to subscribe to that oath. That custom continues today as part of the graduation ceremony of most medical schools in the United States.

Like most professional codes, business trade agreements, and labor union contracts, the Hippocratic Oath was full of fine ideals for protecting the patient: "I will use my power to help the sick to the best of my ability and judgment. . . . Whenever I go into a house, I will go to help the sick and never with the intention of doing harm or injury. . . ." and so on.

But it also contains a few sleepers. Consider this one: "I will hand on precepts, lectures and all other learning to my sons, to those of my teachers and to those pupils duly apprenticed and sworn, and to none others." Today we would call that the prelude to a closed shop.

Or listen to this one referring to patients suffering from the agonizing disease of kidney or bladder stones: "I will not cut, even for the stone, but I will leave such procedures to the practitioners of that craft," [1] a nice market-sharing agreement between physicians and surgeons.

Hippocrates, we conjecture, must turn in his grave when a new class of medical men takes that oath. He is supposed to have taught everyone who demonstrated the interest and paid his tuition. He would presumably have objected strongly to the kind of restrictive practices that physicians all over the world have adopted from that time to this in order to protect themselves against competition.

The American Medical Association is seldom regarded as a labor union. And it is much more than the ordinary labor union. It renders important services to its members and to the medical profession as a whole. However, it is also a labor union, and in our judgment has been one of the most successful unions in the country. For decades it kept down the number of physicians, kept up the costs of medical care, and prevented competition with "duly apprenticed and sworn" physicians by people from outside the profession—all, of course, in the name of helping the patient. At this point in this book, it hardly needs repeating that the leaders of medicine have been sincere in their belief that restricting entry into medicine would help the patient. By this time we are familiar with the capacity that all of us have to believe that what is in our interest is in the social interest.

As government has come to play a larger role in medicine, and to finance a larger share of medical costs, the power of the American Medical Association has declined. Another monopolistic group, government bureaucrats, has been replacing it. We believe that this result has been brought on partly by the actions of organized medicine itself.

These developments in medicine are important and may have far-reaching implications for the kind and cost of health care that will be available to us in the future. However, this chapter is about labor, not medicine, so we shall refer only to those aspects of medical economics that illustrate principles applicable to all labor union activity. We shall put to one side other important, and fascinating, questions about current developments in the organization of health care.

Who Benefits?

Physicians are among the most highly paid workers in the United States. That status is not exceptional for persons who have benefited from labor unions. Despite the image often conveyed that labor unions protect low-paid workers against exploitation by employers, the reality is very different. The unions that have been most successful invariably cover workers who are in occupations that require skill and would be relatively highly paid with or without unions. These unions simply make high pay still higher.

For example, airline pilots in the United States received an annual salary, for a three-day week, that averaged $50,000 a year in 1976 and has risen considerably since. In a study entitled *The Airline Pilots*, George Hopkins writes, "Today's incredibly high pilot salaries result less from the responsibility pilots bear or the technical skill they possess than from the protected position they have achieved through a union." [2]

The oldest traditional unions in the United States are the craft unions—carpenters, plumbers, plasterers, and the like—again workers who are highly skilled and highly paid. More recently, the fastest growing unions—and indeed almost the only ones that have grown at all—are unions of government workers, including schoolteachers, policemen, sanitation workers, and every other variety of government employee. The municipal unions in New York City have demonstrated their strength by helping to bring that city to the verge of bankruptcy.

Schoolteachers. and municipal employees illustrate a general principle that is clearly exemplified in Great Britain. Their unions do not deal directly with the taxpayers who pay their members' salaries. They deal with government officials. The looser the connection between taxpayers and the officials the unions deal with, the greater the tendency for officials and the unions to gang up at the expense of the taxpayer—another example of what happens when some people spend other people's money on still other people. That is why municipal unions are stronger in large cities such as New York than in small cities, why unions of schoolteachers have become more powerful as control over the conduct of schools and over educational expenditures has become more centralized, further removed from the local community.

In Great Britain the government has nationalized many more industries than in the United States—including coal mining, public utilities, telephones, hospitals. And labor unions in Britain have generally been strongest, and labor problems most serious, in the nationalized industries. The same principle is reflected in the strength of the U.S. postal unions.

Given that members of strong unions are highly paid, the obvious question is: are they highly paid because their unions are strong, or are their unions strong because they are highly paid? Defenders of the unions claim that the high pay of their members is a tribute to the strength of union organization, and that if only all workers were members of unions, all workers would be highly paid.

The situation is, however, much more complex. Unions of highly skilled workers have unquestionably been able to raise the wages of their members; however, people who would in any event be highly paid are in a favorable position to form strong unions. Moreover, the ability of unions to raise the wages of some workers does not mean that universal unionism could raise the wages of all workers. On the contrary, and this is a fundamental source of misunderstanding, *the gains that strong unions win for their members are primarily at the expense of other workers.*

The key to understanding the situation is the most elementary principle of economics: the law of demand—the higher the price of anything, the less of it people will be willing to buy. Make labor of any kind more expensive and the number of jobs of that kind will be fewer. Make carpenters more expensive, and fewer houses than otherwise will be built, and those houses that are built will tend to use materials and methods requiring less carpentry. Raise the wage of airline pilots, and air travel will become more expensive. Fewer people will fly, and there will be fewer jobs for airline pilots. Alternatively, reduce the number of carpenters or pilots, and they will command higher wages. Keep down the number of physicians, and they will be able to charge higher fees. A successful union reduces the number of jobs available of the kind it controls. As a result, some people who would like to get such jobs at the union wage cannot do so. They are forced to look elsewhere. A greater supply of workers for other jobs drives down the wages paid for those jobs. Universal unionization would not alter the situation. It could mean higher wages for the persons who get jobs, along with more unemployment for others. More likely, it would mean strong unions and weak unions, with members of the strong unions getting higher wages, as they do now, at the expense of members of weak unions.

Union leaders always talk about getting higher wages at the expense of profits. That is impossible: profits simply aren't big enough. About 80 percent of the total national income of the United States currently goes to pay the wages, salaries, and fringe benefits of workers. More than half of the rest goes to pay rent and interest on loans. Corporate profits—which is what union leaders always point to—total less than 10 percent of national income. And that is before taxes. After taxes, corporate profits are something like 6 percent of the national income. That hardly provides much leeway to finance higher wages, even if all profits were absorbed. And that would kill the goose that lays the golden eggs. The small margin of profit provides the incentive for investment in factories and machines, and for developing new products and methods. This investment, these innovations, have, over the years, raised the productivity of the worker and provided the wherewithal for higher and higher wages.

Higher wages to one group of workers must come primarily from other workers. Nearly thirty years ago one of us estimated that on the average about 10 to 15 percent of the workers in this country had been able through unions or their equivalent, such as the American Medical Association, to raise their wages 10 to 15 percent above what they otherwise would have been, at the cost of reducing the wages earned by the other 85 to 90 percent by some 4 percent below what they otherwise would have been.

More recent studies indicate that this remains roughly the order of magnitude of the effect of unions.[3] Higher wages for high-paid workers, lower wages for low-paid workers.

All of us, including the highly unionized, have indirectly been harmed as consumers by the effect of high union wages on the prices of consumer goods. Houses are unnecessarily expensive for everyone, including the carpenters. Workers have been prevented by unions from using their skills to produce the most highly valued items; they have been forced to resort to activities where their productivity is less. The total basket of goods available to all of us is smaller than it would have been.

The Source of Union Power

How can unions raise the wages of their members? What is the basic source of their power? The answer is: the ability to keep down the number of jobs available, or equivalently, to keep down the number of persons available for a class of jobs. Unions have been able to keep down the number of jobs by enforcing a high wage rate, generally with assistance from government. They have been able to keep down the number of persons available, primarily through licensure, again with government aid. They have occasionally gained power by colluding with employers to enforce a monopoly of the product their members help to produce.

Enforcing a high wage rate. If, somehow or other, a union can assure that no contractor will pay less than, say, $15 an hour for a plumber or a carpenter, that will reduce the number of jobs that will be offered. Of course, it will also increase the number of persons who would like to get jobs.

Suppose for the moment that the high wage rate can be enforced. There must then be some way to ration the limited number of lucrative jobs among the persons seeking them. Numerous devices have been adopted: nepotism—to keep the jobs in the family; seniority and apprenticeship rules; featherbedding—to spread the work around; and simple corruption. The stakes are high, so the devices used are a sensitive matter in union affairs. Some unions will not permit seniority provisions to be discussed in open meetings because that always leads to fistfights. Kickbacks to union officials to secure preference for jobs are a common form of corruption. The heavily criticized racial discrimination by unions is still another device for rationing jobs. If there is a surplus of applicants for a limited number of jobs to be rationed, any device to select the ones who get the jobs is bound to be arbitrary. Appeals to prejudice and similar irrational considerations often have great support among the "ins" as a way of deciding whom to keep out. Racial and religious discrimination have entered also into admissions to medical schools and for the same reason: a surplus of acceptable applicants and the need to ration places among them.

To return to the wage rate, how can a union enforce a high wage rate? One way is violence or the threat of violence: threatening to destroy the property of employers, or to beat them up if they employ nonunion workers or if they pay union members less than the union-specified rate; or to beat up workers, or destroy their property, if they agree to work for a lower wage. That is the reason union wage arrangements and negotiations have so often been accompanied by violence.

An easier way is to get the government to help. That is the reason union headquarters are clustered around Capitol Hill in Washington, why they devote so much money and attention to politics. In his study of the airline pilots' union, Hopkins notes that "the union secured enough federal protective legislation to make the professional airline pilots practically a ward of the state." [4]

A major form of government assistance to construction unions is the Davis-Bacon Act, a federal law that requires all contractors who work on a contract in excess of $2,000 to which the U.S. government or the District of Columbia is a party to pay wage rates no less than those "prevailing for the corresponding classes of laborers and mechanics" in the neighborhood in question, as "determined by the Secretary of Labor." In practice the "prevailing" rates have been ruled to be union wage rates in "an overwhelming proportion of wage determinations . . . regardless of area or type of construction." [5] The reach of the act has been extended by the incorporation of its prevailing wage requirement in numerous other laws for federally assisted projects, and by similar laws in thirty-five states (as of 1971) covering state construction expenditures.[6] The effect of these acts is that the government enforces union wage rates for much of construction activity.

Even the use of violence implicitly involves government support. A generally favorable public attitude toward labor unions has led the authorities to tolerate behavior in the course of labor disputes that they would never tolerate under other circumstances. If someone's car gets overturned in the course of a labor dispute, or if plant, store, or home windows get smashed, or if people even get beaten up and seriously injured, the perpetrators are less likely to pay a fine, let alone go to jail, than if the same incident occurred under other circumstances.

Another set of government measures enforcing wage rates are minimum wage laws. These laws are defended as a way to help low-income people. In fact, they hurt low-income people. The source of pressure for them is demonstrated by the people who testify before Congress in favor of a higher minimum wage. They are not representatives of the poor people. They are mostly representatives of organized labor, of the AFL-CIO and other labor organizations. No member of their unions works for a wage anywhere close to the legal minimum. Despite all the rhetoric about helping the poor, they favor an ever higher minimum wage as a way to protect the members of their unions from competition.

The minimum wage law requires employers to discriminate against persons with low skills. No one describes it that way, but that is in fact what it is. Take a poorly educated teenager with little skill whose services are worth, say, only $2.00 an hour. He or she might be eager to work for that wage in order to acquire greater skills that would permit a better job. The law says that such a person may be hired only if the employer is willing to pay him or her (in 1979) $2.90 an hour. Unless an employer is willing to add 90 cents in charity to the $2.00 that the person's services are worth, the teenager will not be employed. It has always been a mystery to us why a young person is better off unemployed from a job that would pay $2.90 an hour than employed at a job that does pay $2.00 an hour.

The high rate of unemployment among teenagers, and especially black teenagers, is both a scandal and a serious source of social unrest. Yet it is largely a result of minimum wage laws. At the end of World War II the minimum wage was 40 cents an hour. Wartime inflation had made that so low in real terms as to be unimportant. The minimum wage was then raised sharply to 75 cents in 1950, to $1.00 in 1956. In the early fifties the unemployment rate for teenagers averaged 10 percent compared with about 4 percent for all workers—moderately higher, as one would expect for a group just entering the labor force. The unemployment rates for white and black teenagers were roughly equal. After minimum wage rates were raised sharply, the unemployment rate shot up for both white and black teenagers. Even more significant, an unemployment gap opened between the rates for white and black teenagers. Currently, the unemployment rate runs around 15 to 20 percent for white teenagers; 35 to 45 percent for black teenagers.[7] We regard the minimum wage rate as one of the most, if not the most, antiblack laws on the statute books. The government first provides schools in which many young people, disproportionately black, are educated so poorly that they do not have the skills that would enable them to get good wages. It then penalizes them a second time by preventing them from offering to work for low wages as a means of inducing employers to give them on-the-job training. All this is in the name of helping the poor.

Restricting numbers. An alternative to enforcing a wage rate is to restrict directly the number who may pursue an occupation. That technique is particularly attractive when there are many employers—so that enforcing a wage rate is difficult. Medicine is an excellent example, since much of the activity of organized medicine has been directed toward restricting the number of physicians in practice.

Success in restricting numbers, as in enforcing a wage rate, generally requires the assistance of the government. In medicine the key has been the licensure of physicians—that is, the requirement that in order for any individual to "practice medicine," he must be licensed by the state. Needless to say, only physicians are likely to be regarded as competent to judge the qualifications of potential physicians, so licensing boards in the various states (in the United States licensure is under the jurisdiction of the state, not the federal government) are typically composed wholly of physicians or dominated by physicians, who in turn have generally been members of the AMA.

The boards, or the state legislatures, have specified conditions for the granting of licenses that in effect give the AMA the power to influence the number of persons admitted to practice. They have required lengthy training, almost always graduation from an "approved" school, generally internship in an "approved" hospital. By no accident, the list of "approved" schools and hospitals is generally identical with the list issued by the Council on Medical Education and Hospitals of the American Medical Association. No school can be established or, if established, long continue unless it can get the approval of the AMA Council on Medical Education. That has at times required limiting the number of persons admitted in accordance with the council's advice.

Striking evidence of the power of organized medicine to restrict entry was provided during the depression of the 1930s when the economic pressure was particularly great. Despite a flood of highly trained refugees from Germany and Austria—at the time centers of advanced medicine—the number of foreign-trained physicians admitted to practice in the United States in the five years after Hitler came to power was no larger than in the preceding five years.[8]

Licensure is widely used to restrict entry, particularly for occupations like medicine that have many individual practitioners dealing with a large number of individual customers. As in medicine, the boards that administer the licensure provisions are composed primarily of members of the occupation licensed—whether they be dentists, lawyers, cosmetologists, airline pilots, plumbers, or morticians. There is no occupation so remote that an attempt has not been made to restrict its practice by licensure. According to the chairman of the Federal Trade Commission: "At a recent session of one state legislature, occupational groups advanced bills to license themselves as auctioneers, well-diggers, home improvement contractors, pet groomers, electrologists, sex therapists, data processors, appraisers, and TV repairers. Hawaii licenses tattoo artists. New Hampshire licenses lightning-rod salesmen." [9] The *justification* offered is always the same: to protect the consumer. However, the *reason* is demonstrated by observing who lobbies at the state legislature for the imposition or strengthening of licensure. The lobbyists are invariably representatives of the occupation in question rather than of the customers. True enough, plumbers presumably know better than anyone else what their customers need to be protected against. However, it is hard to regard altruistic concern for their customers as the primary motive behind their determined efforts to get legal power to decide who may be a plumber.

To reinforce the restriction on numbers, organized occupational groups persistently strive to have the practice of their occupation legally defined as broadly as possible in order to increase the demand for the services of licensed practitioners.

One effect of restricting entry into occupations through licensure is to create new disciplines: in medicine, osteopathy and chiropractic are examples. Each of these, in turn, has resorted to licensure to try to restrict its numbers. The AMA has engaged in extensive litigation charging chiropractors and osteopaths with the unlicensed practice of medicine, in an attempt to restrict them to as narrow an area as possible. Chiropractors and osteopaths in turn charge other practitioners with the unlicensed practice of chiropractic and osteopathy.

A recent development in health care, arising partly out of new, sophisticated portable equipment, has been the development of services in various communities to bring prompt aid in emergencies. These services are sometimes organized by the city or a city agency, sometimes by a strictly private enterprise, and are manned primarily by paramedics rather than licensed physicians.

Joe Dolphin, the owner of one such private enterprise organization attached to a fire department in southern California, described its effectiveness as follows:

> In one district of California that we serve, which is a county which is populated to the extent of five hundred and eighty thousand people, before the introduction of paramedics, less than one percent of the patients that suffered a cardiac arrest where their heart stopped lived through their hospital stay and were released from the hospital. With the introduction of paramedics, just in the first six months of opera-

tion, twenty-three percent of the people whose heart stops are successfully resurrected and are released from the hospital and go back to productive work in society. We think that's pretty amazing. We think the facts speak for themselves. However, relating that to the medical community is sometimes very difficult. They have ideas of their own.

More generally, jurisdictional disputes—what activities are reserved to what occupation—are among the most frequent sources of labor stoppages. An amusing example was a reporter for a radio station who came to interview one of us. He emphasized that the interview had to be short enough to fit on one side of the cassette in his cassette recorder. Turning over the cassette was reserved to a member of the electricians' union. If, he said, he turned it over himself, the cassette would be erased when he returned to the station, and the interview lost. Exactly the same behavior as the medical profession's opposition to paramedics, and motivated by the same objective: to increase the demand for the services of a particular group.

Collusion between unions and employers. Unions have sometimes gained power by helping business enterprises combine to fix prices or share markets, activities that are illegal for business under the antitrust laws.

The most important historical case was in coal mining in the 1930s. The two Guffey coal acts were attempts to provide legal support for a price-fixing cartel of coal mine operators. When, in the mid-thirties, the first of the acts was declared unconstitutional, John L. Lewis and the United Mine Workers that he headed stepped into the breach. By calling strikes or work stoppages whenever the amount of coal above the ground got so large as to threaten to force down prices, Lewis controlled output and thereby prices with the unspoken cooperation of the industry. As the vice-president of a coal company put it in 1938, "They [the United Mine Workers] have done a lot to stabilize the bituminous coal industry and have endeavored to have it operate on a profitable basis, in fact though one dislikes to admit it their efforts along that line have in the main . . . been a bit more efficacious . . . than the endeavors of coal operators themselves." [10]

The gains were divided between the operators and the miners. The miners were granted high wage rates, which of course meant greater mechanization and fewer miners employed. Lewis recognized this effect explicitly and was more than prepared to accept it—regarding higher wages for miners employed as ample compensation for a reduction in the number employed, provided those employed were all members of his union.

The miners' union could play this role because unions are exempt from the Sherman Anti-Trust Act. Unions that have taken advantage of this exemption are better interpreted as enterprises selling the services of cartelizing an industry than as labor organizations. The Teamsters' Union is perhaps the most notable. There is a story, perhaps apocryphal, about David Beck, the head of the Teamsters' Union before James Hoffa (both of whom ultimately went to jail). When Beck was negotiating with breweries in the state of Washington about wages for drivers of brewery trucks, he was told that the wages he was asking were not feasible because "eastern beer" would undercut local beer. He asked what the price of eastern beer would have to be to permit the wage he demanded. A figure, X dollars a case, was named, and he supposedly replied, "From now on, eastern beer will be X dollars a case."

Labor unions can and often do provide useful services for their members—negotiating the terms of their employment, representing them with respect to grievances, giving them a feeling of belonging and participating in a group activity, among others. As believers in freedom, we favor the fullest opportunity for voluntary organization of labor unions to perform whatever services their members wish, and are willing to pay for, provided they respect the rights of others and refrain from using force.

However, unions and comparable groups such as the professional associations have not relied on strictly voluntary activities and membership with respect to their major proclaimed objective—improving the wages of their members. They have succeeded in getting government to grant them special privileges and immunities, which have enabled them to benefit some of their members and officials at the expense of other workers and all consumers. In the main, the persons benefited have had decidedly higher incomes than the persons harmed.

GOVERNMENT

In addition to protecting union members, government has adopted a host of laws intended to protect workers in general: laws that provide for workmen's compensation, prohibit child labor, set minimum wages and maximum hours of labor, establish commissions to assure fair employment practices, promote affirmative action, establish the federal Office of Safety and Health Administration to regulate employment practices, and others too numerous to list.

Some measures have had a favorable effect on conditions of work. Most, like workmen's compensation and child labor laws, simply embodied in law practices that had already become common in the private market, perhaps extending them somewhat to fringe areas. Others, you will not be surprised to learn, have been a mixed blessing. They have provided a source of power for particular unions or employers, and a source of jobs for bureaucrats, while reducing the opportunities and incomes of the ordinary worker. OSHA is a prime example—a bureaucratic nightmare that has produced an outpouring of complaints on all sides. As a recent joke has it: How many Americans does it take to screw in a light bulb? Answer: Five; one to screw in the bulb, four to fill out the environmental impact and OSHA reports.

Government does protect one class of workers very well, namely, those employed by government.

Montgomery County, Maryland, a half-hour's drive from Washington, D.C., is the home of many senior civil servants. It also has the highest average family income of any county in the United States. One out of every four employed persons in Montgomery County works for the federal government. They have job security and salaries linked to the cost of living. At retirement they receive civil service pensions also linked to the cost of living and independent of Social Security. Many manage to qualify for Social Security as well, becoming what are known as double dippers.

Many, perhaps most, of their neighbors in Montgomery County also have some connection with the federal government—as congressmen, lobbyists, top executives of corporations with government contracts. Like other bedroom communities around Washington, Montgomery County has been growing rapidly. Government has become a highly dependable growth industry in recent decades.

All civil servants, even at low levels, are well protected by the government. According to most studies, their salaries average higher than comparable private salaries and are protected against inflation. They get generous fringe benefits and have an almost incredible degree of job security.

As a *Wall Street Journal* story put it:

> As the [Civil Service] regulations have ballooned to fill 21 volumes some five feet thick, government managers have found it increasingly difficult to fire employees. At the same time, promotions and merit pay raises have become almost automatic. The result is a bureaucracy nearly devoid of incentives and largely beyond anyone's control. . . .
>
> Of the one million people eligible last year for merit raises, only 600 didn't receive them. Almost no one is fired; less than 1% of federal workers lost their jobs last year.[11]

To cite one specific case, in January 1975 a typist in the Environmental Protection Agency was so consistently late for work that her supervisor demanded she be fired. It took nineteen months to do it—and it takes a twenty-one-foot-long sheet to list the steps that had to be gone through to satisfy all the rules and all the management and union agreements.

The process involved the employee's supervisor, the supervisor's deputy director and director, the director of personnel operations, the agency's branch chief, an employee relations specialist, a second employee relations specialist, a special office of investigations, and the director of the office of investigations. Needless to say, this veritable telephone directory of officials was paid for with taxpayers' money.

At state and local levels the situation varies from place to place. In many states and in large cities such as New York, Chicago, and San Francisco, the situation is either the same as or more extreme than in the federal government. New York City was brought to its present state of virtual bankruptcy largely by rapid increases in the wages of municipal employees and, perhaps even more, by the granting of generous pensions at early retirement ages. In states with big cities, representatives of public employees are often the major special interest group in the state legislature.

NO ONE

Two classes of workers are not protected by anyone: workers who have only one possible employer, and workers who have no possible employer.

The individuals who effectively have only one possible employer tend to be highly paid people whose skills are so rare and valuable that only one employer is big enough or well enough situated to take full advantage of them.

The standard textbook example when we studied economics in the 1930s was the great baseball hero Babe Ruth. The "Sultan of Swat," as the home run king was nicknamed, was by far the most popular baseball player of his time. He could fill any stadium in either of the major leagues. The New York Yankees happened to have the largest stadium of any baseball club, so it could afford to pay him more than any other club. As a result, the Yankees were effectively his only possible employer. That doesn't mean, of course, that Babe Ruth didn't succeed in commanding a high salary, but it did mean that he had no one to protect him; he had to bargain with the Yankees, using the threat of not playing for them as his only weapon.

Individuals who have no choice among employers are mostly the victims of government measures. One class has already been mentioned: those who are rendered unemployed by legal minimum wages. As noted earlier, many of them are double victims of government measures: poor schooling plus high minimum wages that prevent them from getting on-the-job training.

Persons on relief or public assistance are in a somewhat similar position. It is to their advantage to take employment only if they can earn enough to make up for the loss of their welfare or other public assistance. There may be no employer to whom their services are worth that much. That is true also of persons on Social Security and less than seventy-two years old. They lose their Social Security benefits if they earn more than a modest amount. That is the major reason why the fraction of persons over sixty-five years old who are in the labor force has decreased so sharply in recent decades: for males, from 45 percent in 1950 to 20 percent in 1977.

OTHER EMPLOYERS

The most reliable and effective protection for most workers is provided by the existence of many employers. As we have seen, a person who has only one possible employer has little or no protection. The employers who protect a worker are those who would like to hire him. Their demand for his services makes it in the self-interest of his own employer to pay him the full value of his work. If his own employer doesn't, someone else may be ready to do so. Competition for his services—that is the worker's real protection.

Of course, competition by other employers is sometimes strong, sometimes weak. There is much friction and ignorance about opportunities. It may be costly for employers to locate desirable employees, and for employees to locate desirable employers. This is an imperfect world, so competition does not provide complete protection. However, competition is the best, or, what is the same thing, the least bad, protection for the largest number of workers that has yet been found or devised.

The role of competition is a feature of the free market that we have encountered time and again. A worker is protected from his employer by the existence of other employers for whom he can go to work. An employer is protected from exploitation by his employees by the existence of other workers whom he can hire. The consumer is protected from exploitation by a given seller by the existence of other sellers from whom he can buy.

Why do we have poor postal service? Poor long-distance train service? Poor schools? Because in each case there is essentially only one place we can get the service.

CONCLUSION

When unions get higher wages for their members by restricting entry into an occupation, those higher wages are at the expense of other workers who find their opportunities reduced. When government pays its employees higher wages, those higher wages are at the expense of the taxpayer. But when workers get higher wages and better working conditions through the free market, when they get raises by firms competing with one another for the best workers, by workers competing with one another for the best jobs, those higher wages are at nobody's expense. They can only come from higher productivity, greater capital investment, more widely diffused skills. The whole pie is bigger—there's more for the worker, but there's also more for the employer, the investor, the consumer, and even the tax collector.

That's the way a free market system distributes the fruits of economic progress among all the people. That's the secret of the enormous improvement in the conditions of the working person over the past two centuries.

CHAPTER 9

The Cure
for Inflation

Compare two rectangles of paper of about the same size. One is mostly green on the back side and has a picture of Abraham Lincoln on the front side, which also has the number 5 on each of its corners and some printing. You can exchange this piece of paper for some quantity of food, clothing, or other goods. People will willingly make the trade.

The other piece of paper, perhaps cut from a glossy magazine, may also have a picture, some numbers, and some printing on its face. It may also be colored green on its back. Yet it is fit only to light the fire.

Whence the difference? The printing on the $5 bill gives no answer. It simply says, "FEDERAL RESERVE NOTE / THE UNITED STATES OF AMERICA / FIVE DOLLARS" and, in smaller print, "THIS NOTE IS LEGAL TENDER FOR ALL DEBTS, PUBLIC AND PRIVATE." Until not very many years ago, the words "WILL PROMISE TO PAY" were included between "THE UNITED STATES OF AMERICA" and "FIVE DOLLARS." That seemed to explain the difference between the two pieces of paper. But it meant only that if you had gone to a Federal Reserve Bank and asked a teller to redeem the promise, he would have given you five identical pieces of paper except that the number 1 took the place of the number 5 and George Washington's picture the place of Abraham Lincoln's. If you had then asked the teller to pay the $1 promised by one of these pieces of paper, he would have given you coins which, if you had melted them down (despite its being illegal to do so), would have sold for less than $1 as metal. The present wording is at least more candid if equally unrevealing. The legal-tender quality means that the government will accept the pieces of paper in discharge of debts and taxes due to itself, and that the courts will regard them as discharging debts stated in dollars. Why should they also be accepted by private persons in private transactions in exchange for goods and services?

The short answer is that each person accepts them because he is confident that others will. The pieces of green paper have value because everybody thinks they have value. Everybody thinks they have value because in his experience they have had value. The United States could not operate at more than a small fraction of its present level of productivity without a common and widely accepted medium of exchange (or at most a small number of such media); yet the existence of a common and widely accepted medium of exchange rests on a convention that owes its existence to the mutual acceptance of what, from one point of view, is a fiction.

The convention or the fiction is no fragile thing. On the contrary, the value of having a common money is so great that people will stick to the fiction even under extreme provocation— whence, as we shall see, comes part of the gain that issuers of the money can derive from inflation and hence the temptation to inflate. But neither is the fiction indestructible: the phrase "not worth a Continental" is a reminder of how that fiction was destroyed for the Continental currency issued in excessive amount by the U.S. Continental Congress to finance the American Revolution.

Though the value of money rests on a fiction, money serves an extraordinarily useful economic function. Yet it is also a veil. The "real" forces that determine the wealth of a nation are the capacities of its citizens, their industry and ingenuity, the resources at their command, their mode of economic and political organization, and the like. As John Stuart Mill wrote more than a century ago: "There cannot, in short, be intrinsically a more insignificant thing, in the economy of society, than money; except in the character of a contrivance for sparing time and labour. It is a machine for doing quickly and commodiously, what would be done, though less quickly and commodiously, without it: and like many other kinds of machinery, it only exerts a distinct and independent influence of its own when it gets out of order." [1]

Perfectly true, as a description of the role of money, provided we recognize that society possesses hardly any other contrivance that can do more damage when it gets out of order.

We have already discussed one example: the Great Depression, when money got out of order through too sharp a reduction in its quantity. This chapter discusses the opposite and more common way in which money has gotten out of order—through too sharp an increase in quantity.

VARIETIES OF MONEY

An amazing variety of items has been used as money at one time or anoher. The word "pecuniary" comes from the Latin *pecus*, meaning "cattle," one of the many things that have been used as money. Others include salt, silk, furs, dried fish, even feathers, and, on the Pacific island of Yap, stones. Cowrie shells and beads have been the most widely used forms of primitive money. Metals—gold, silver, copper, iron, tin—have been the most widely used forms among more advanced economies before the victory of paper and the bookkeeper's pen.

The one thing all the items used as money have had in common is their acceptance, in the particular place and time, in return for other goods and services in the faith that others would likewise accept them.

The "wampum" that the early settlers of America used in trade with Indians was a form of shell, analogous to the cowrie shells used in Africa and Asia. A most interesting and instructive money used in the American colonies was the tobacco money of Virginia, Maryland, and North Carolina: "The first law passed by the first General Assembly of Virginia, July 31, 1619 [twelve years after Captain John Smith landed and established at Jamestown the first permanent settlement in the New World], was in reference to tobacco. It fixed the price of that staple 'at three shillings the beste, and the second sorte at 18d. the pounde.' . . . Tobacco was already the local currency." [2]

At various periods tobacco was declared the only legal currency. It remained a basic money of Virginia and its neighboring colonies for close to two centuries, until well after the American Revolution. It was the money that the colonists used to buy food, clothing, to pay taxes—even to pay for a bride: "The Rev. Mr. Weems, a Virginian writer, intimates that it would have done a man's heart good to see the gallant young Virginians hastening to the waterside when a vessel arrived from London, each carrying a bundle of the best tobacco under his arm, and taking back with him a beautiful and virtuous young wife." [3] And another writer, quoting this passage, goes on to remark, "They must have been stalwart, as well as gallant, to hasten with a roll of tobacco weighing 100 to 150 pounds under the arm." [4]

As money goes, so tobacco went. The original price set on it in terms of English money was higher than the cost of growing it, so planters set to with a will and produced more and more. In this case, the money supply grew literally as well as figuratively. As always happens when the quantity of money increases more rapidly than the quantity of goods and services available for purchase, there was inflation. Prices of other things in terms of tobacco rose drastically. Before the inflation ended about half a century later, prices in terms of tobacco had risen fortyfold.

The growers of tobacco were most unhappy about the inflation. Higher prices of other things in terms of tobacco meant that tobacco would command less of those other things. The price of money in terms of goods is the reciprocal of the price of goods in terms of money. Naturally, tobacco growers turned to government for help. One law after another was passed prohibiting certain classes of people from growing tobacco; providing for destroying part of the crop; prohibiting the planting of tobacco for one year. All to no avail. Finally, people took matters into their own hands, banded together, and went around the countryside destroying tobacco plants: "The evil reached such proportions that in April, 1684, the Assembly passed a law declaring that these malefactors had passed beyond the bounds of riot, and that their aim was the subversion of the government. It was enacted that if any persons to the number of eight or more should go about destroying tobacco plants, they should be adjudged traitors and suffer death." [5]

The tobacco currency vividly illustrates one of the oldest laws in economics: Gresham's Law, "Bad money drives out good." The grower of tobacco, who had to pay taxes or other obligations fixed in terms of tobacco, understandably used the poorest quality tobacco to discharge obligations and retained the best quality for export in return for "hard" money, i.e., British sterling. As a result, only poor quality tobacco tended to circulate as money. Every device of human ingenuity was used to make tobacco appear higher in quality than it was: "Maryland in 1698 found it necessary to legislate against the fraud of packing trash in hogsheads that contained good tobacco on top. Virginia adopted a similar measure in 1705, but apparently it did not offer relief." [6]

The quality problem was somewhat alleviated when "[i]n 1727 tobacco notes were legalized. These were in the nature of certificates of deposit issued by the inspectors. They were declared by law current and payable for all tobacco debts within the warehouse district where they were issued." [7] Despite numerous abuses of the system, "[s]uch receipts performed the office of currency right to the eve of the 19th century." [8]

That was not the last use of tobacco as money. During World War II cigarettes were widely used as a medium of exchange in German and Japanese prison camps. After World War II cigarettes were widely used as money in Germany during the period when the occupation authorities enforced ceilings on prices in legal currency that were well below the levels that would have cleared the market. The result was to destroy the usefulness of the legal money. People resorted to barter and to the use of cigarettes as a medium of exchange for small transactions, and cognac for large ones—by all odds the most liquid currency of which we have record. Ludwig Erhard's monetary reform ended that instructive—and destructive—episode. [9]

The general principles illustrated by tobacco money in Virginia remain relevant in the modern era, though paper money issued by government and bookkeeping entries called deposits have replaced commodities or warehouse receipts for commodities as the basic money of the society.

It remains as true now as it was then that a more rapid increase in the quantity of money than in the quantity of goods and services available for purchase will produce inflation, raising prices in terms of that money. It does not matter why the quantity of money increases. In Virginia the quantity of tobacco money grew and produced an inflation of prices in terms of tobacco because the cost of producing tobacco in terms of labor and other resources fell drastically. In Europe in the Middle Ages, silver and gold were the dominant money, and inflation of prices in terms of gold and silver occurred because precious metals from Mexico and South America flooded Europe via Spain. In the mid-nineteenth century inflation of prices in terms of gold occurred around the world because of gold discoveries in California and Australia; later, from the 1890s to 1914, because of the successful commercial application of the cyanide process to the extraction of gold from low-grade ore, primarily in South Africa.

Today, when the commonly accepted media of exchange have no relation to any commodity, the quantity of money is determined in every major country by government. Government and the government alone is responsible for any rapid increase in the quantity of money. That very fact has been the major source of confusion about the cause and the cure of inflation.

THE PROXIMATE CAUSE OF INFLATION

Inflation is a disease, a dangerous and sometimes fatal disease, a disease that if not checked in time can destroy a society. Examples abound. Hyperinflations in Russia and Germany after World War I—when prices sometimes doubled and more than doubled from one day to the next—prepared the ground for communism in the one country and nazism in the other. The hyperinflation in China after World War II eased Chairman Mao's defeat of Chiang Kai-shek. Inflation in Brazil, where it reached about 100 percent a year in 1954, brought military government. A far more extreme inflation contributed to the overthrow of Allende in Chile in 1973 and of Isabel Perón in Argentina in 1976, followed in both countries by the assumption of power by a military junta.

No government is willing to accept responsibility for producing inflation, even in less virulent degree. Government officials always find some excuse—greedy businessmen, grasping trade unions, spendthrift consumers, Arab sheikhs, bad weather, or anything else that seems even remotely plausible. No doubt, businessmen are greedy, trade unions are grasping, consumers are spendthrifts, Arab sheikhs have raised the price of oil, and weather is often bad. All these can produce high prices for individual items; they cannot produce rising prices for goods in general. They can cause temporary ups or downs in the rate of inflation. But they cannot produce continuing inflation for one very simple reason: none of the alleged culprits possesses a printing press on which it can turn out those pieces of paper we carry in our pockets; none can legally authorize a bookkeeper to make entries on ledgers that are the equivalent of those pieces of paper.

Inflation is not a capitalist phenomenon. Yugoslavia, a communist country, has experienced one of the most rapid rates of inflation of any country in Europe; Switzerland, a bastion of capitalism, one of the lowest. Neither is inflation a communist phenomenon. China had little inflation under Mao; Italy, the United Kingdom, Japan, the United States—all largely capitalist countries—have experienced substantial inflation in the past decade. In the modern world, inflation is a printing press phenomenon.

The recognition that substantial inflation is always and everywhere a monetary phenomenon is only the beginning of an understanding of the cause and cure of inflation. The more basic question is, why do modern governments increase the quantity of money too rapidly? Why do they produce inflation when they understand its potential for harm?

Before turning to that question, it is worth dwelling a bit longer on the proposition that inflation is a monetary phenomenon. Despite the importance of that proposition, despite the extensive historical evidence supporting it, it is still widely denied—in large part because of the smoke screen with which governments try to conceal their own responsibility for inflation.

If the quantity of goods and services available for purchase— output, for short—were to increase as rapidly as the quantity of money, prices would tend to be stable. Prices might even fall gradually as higher incomes led people to want to hold a larger fraction of their wealth in the form of money. Inflation occurs when the quantity of money rises appreciably more rapidly than output, and the more rapid the rise in the quantity of money per unit of output, the greater the rate of inflation. There is probably no other proposition in economics that is as well established as this one.

Output is limited by the physical and human resources available and by the improvement in knowledge and capacity to use them. At best, output can grow only fairly slowly. Over the past century, output in the United States grew at the average rate of about 3 percent per year. Even at the height of the rapid growth of Japan after World War II, output grew about 10 percent per year. The quantity of commodity money is subject to similar physical limits, though, as the examples of tobacco, precious metals from the New World, and gold in the nineteenth century illustrate, commodity money has at times grown far more rapidly than output in general. Modern forms of money—paper and bookkeeping entries—are subject to no physical limits. The nominal quantity, that is, the number of dollars, pounds, marks, or other monetary units, can grow at any rate, and at times has grown at fantastic rates.

During the German hyperinflation after World War I, for example, hand-to-hand money grew at the *average* rate of more than 300 percent a *month* for more than a year, and so did prices. During the Hungarian hyperinflation after World War II, hand-to-hand money rose at the *average* rate of more than 12,000 percent per *month* for a year, and prices at the even higher rate of nearly 20,000 percent a month.[10]

During the far more moderate inflation in the United States from 1969 to 1979, the quantity of money rose at the average rate of 9 percent per year and prices at the average rate of 7 percent per year. The difference of two percentage points reflects the 2.8 percent average rate of growth of output over the same decade.

As these examples show, what happens to the quantity of money tends to dwarf what happens to output; hence our reference to inflation as a *monetary* phenomenon, without adding any qualification about output. These examples also show that there is not a precise one-to-one correspondence between the rate of monetary growth and the rate of inflation. However, to our knowledge there is no example in history of a substantial inflation that lasted for more than a brief time that was not accompanied by a roughly correspondingly rapid increase in the quantity of money; and no example of a rapid increase in the quantity of money that was not accompanied by a roughly correspondingly substantial inflation.

A few charts (Figures 1–5) show the persistence of this relation in recent years. The solid line on each chart is the quantity of money per unit of output for the country in question, year by year from 1964 through 1977. The other line is the consumer price index. In order to make the two series comparable, both have been expressed as percentages of their average values over the period as a whole (1964–1977 = 100 for both lines). The two lines necessarily have the same average level, but there is nothing in the arithmetic that requires the two lines to be the same for any single year.

The two lines for the United States on Figure 1 are almost indistinguishable. As the remaining figures show, that is not special to the United States. Though the two lines differ more for some of the other countries than they do for the United States, for every country the two lines are remarkably similar. The different countries experienced very different rates of monetary growth. In every case, that difference was matched by a different rate of inflation. Brazil is the most extreme (Figure 5). It experienced more rapid monetary growth than any of the others, and also more rapid inflation.

Which causes which? Does the quantity of money grow rapidly because prices increase rapidly, or vice versa? One clue is that on most of the charts the number plotted for the quantity of money is for a year ending six months *earlier* than the year to which the matching price index corresponds. More decisive evidence is provided by examination of the institutional arrangements that determine the quantity of money in these countries and by a large number of historical episodes in which it is crystal clear which is cause and which is effect.

One dramatic example comes from the American Civil War. The South financed the war largely by the printing press, in the process producing an inflation that averaged 10 percent a month from October 1861 to March 1864. In an attempt to stem the inflation, the Confederacy enacted a monetary reform: "In May, 1864, the currency reform took hold, and the stock of money was reduced. Dramatically, the general price index dropped . . . in

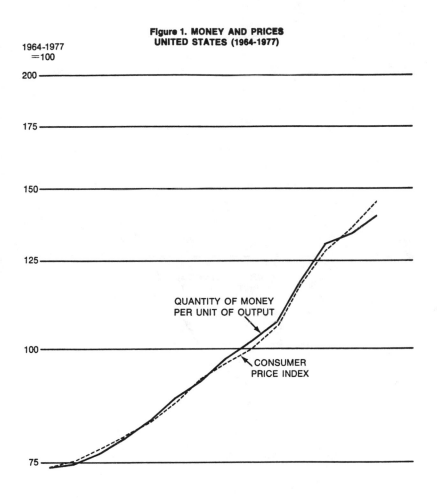

Figure 1. MONEY AND PRICES
UNITED STATES (1964-1977)

1964-1977
=100

VERTICAL SCALE LOGARITHMIC

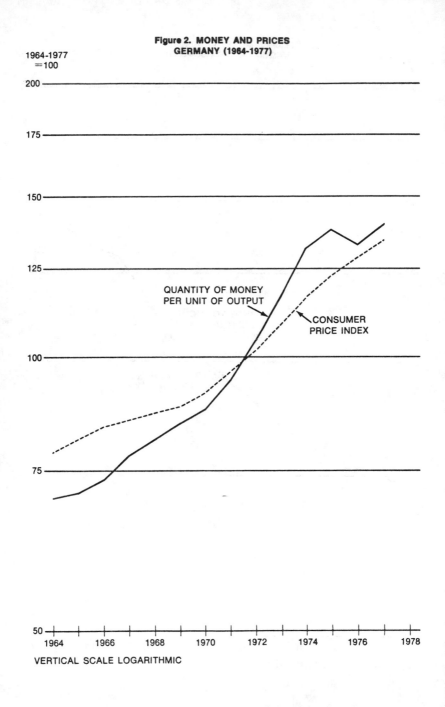

Figure 2. MONEY AND PRICES
GERMANY (1964-1977)

1964-1977
=100

200

175

150

125

QUANTITY OF MONEY
PER UNIT OF OUTPUT

CONSUMER
PRICE INDEX

100

75

50

1964   1966   1968   1970   1972   1974   1976   1978

VERTICAL SCALE LOGARITHMIC

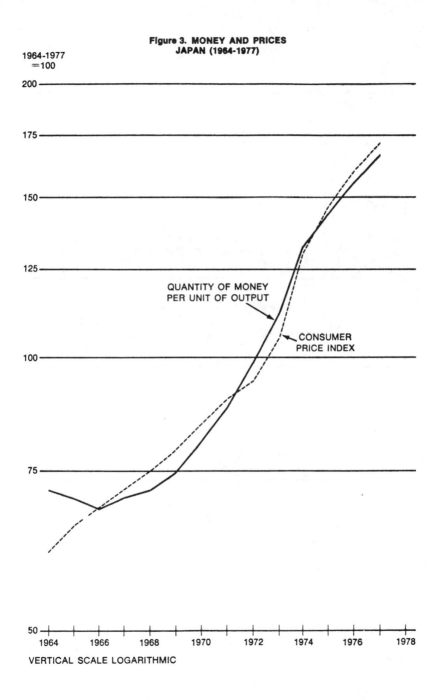

Figure 3. MONEY AND PRICES
JAPAN (1964-1977)

1964-1977
=100

QUANTITY OF MONEY
PER UNIT OF OUTPUT

CONSUMER
PRICE INDEX

VERTICAL SCALE LOGARITHMIC

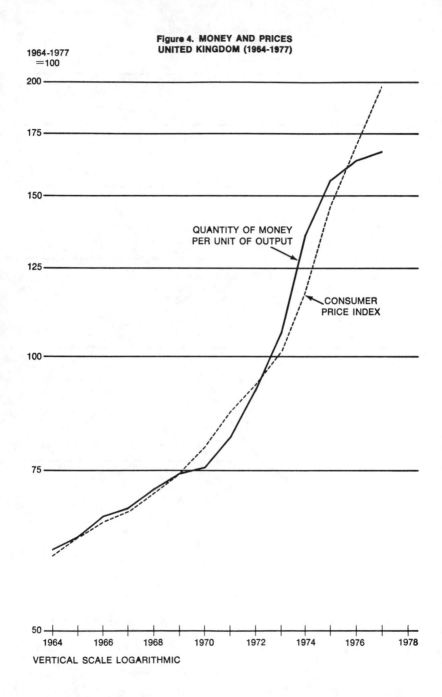

Figure 4. MONEY AND PRICES
UNITED KINGDOM (1964-1977)

1964-1977
=100

QUANTITY OF MONEY
PER UNIT OF OUTPUT

CONSUMER
PRICE INDEX

VERTICAL SCALE LOGARITHMIC

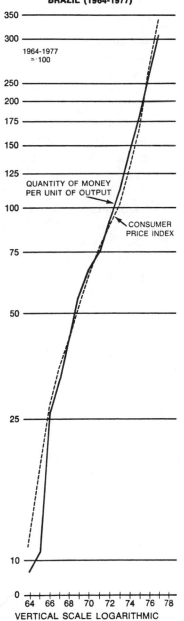

Figure 5. MONEY AND PRICES
BRAZIL (1964-1977)

spite of invading Union armies, the impending military defeat, the reduction in foreign trade, the disorganized government, and the low morale of the Confederate army. Reducing the stock of money had a more significant effect on prices than these powerful forces." [11]

These charts dispose of many widely held explanations of inflation. Unions are a favorite whipping boy. They are accused of using their monopoly power to force up wages, which drive up costs, which drive up prices. But then how is it that the charts for Japan, where unions are of trivial importance, and for Brazil, where they exist only at the sufferance and under the close control of the government, show the same relation as the charts for the United Kingdom, where unions are stronger than in any of the other nations, and for Germany and the United States, where unions have considerable strength? Unions may provide useful services for their members. They may also do a great deal of harm by limiting employment opportunities for others, but they do not produce inflation. Wage increases in excess of increases in productivity are a result of inflation, rather than a cause.

Similarly, businessmen do not cause inflation. The rise in the prices they charge is a result or reflection of other forces. Businessmen are surely no more greedy in countries that have experienced much inflation than in countries that have experienced little, no more greedy at one period than another. Why then is inflation so much greater in some places and at some times than in other places and at other times?

Another favorite explanation of inflation, particularly among government officials seeking to shift blame, is that it is imported from abroad. That explanation was often correct when the currencies of the major countries were linked through a gold standard. Inflation was then an international phenomenon because many countries used the same commodity as money and anything that made the quantity of that commodity money grow more rapidly affected them all. But it clearly is not correct for recent years. If it were, how could the rates of inflation be so different in different countries? Japan and the United Kingdom experienced inflation at the rate of 30 percent or more a year in the early 1970s, when inflation in the United States was around 10 percent and in Germany under 5 percent. Inflation is a worldwide phenomenon in the sense that it occurs in many countries at the same time—just as high goverment spending and large government deficits are worldwide phenomena. But inflation is not an international phenomenon in the sense that each country separately lacks the ability to control its own inflation—just as high government spending and large government deficits are not produced by forces outside each country's control.

Low productivity is another favorite explanation for inflation. Yet consider Brazil. It has experienced one of the most rapid rates of growth in output in the world—and also one of the highest rates of inflation. True enough, what matters for inflation is the quantity of money per unit of output, but as we have noted, as a practical matter, changes in output are dwarfed by changes in the quantity of money. Nothing is more important for the long-run economic welfare of a country than improving productivity. If productivity grows at 3.5 percent per year, output doubles in twenty years; at 5 percent per year, in fourteen years—quite a difference. But productivity is a bit player for inflation; money is center stage.

What about Arab sheikhs and OPEC? They have imposed heavy costs on us. The sharp rise in the price of oil lowered the quantity of goods and services that was available for us to use because we had to export more abroad to pay for oil. The reduction in output raised the price level. But that was a once-for-all effect. It did not produce any longer-lasting effect on the rate of inflation from that higher price level. In the five years after the 1973 oil shock, inflation in both Germany and Japan declined, in Germany from about 7 percent a year to less than 5 percent; in Japan from over 30 percent to less than 5 percent. In the United States inflation peaked a year after the oil shock at about 12 percent, declined to 5 percent in 1976, and then rose to over 13 percent in 1979. Can these very different experiences be explained by an oil shock that was common to all countries? Germany and Japan are 100 percent dependent on imported oil, yet they have done better at cutting inflation than the United States, which is only 50 percent dependent, or than the United Kingdom, which has become a major producer of oil.

We return to our basic proposition. Inflation is primarily a *monetary phenomenon*, produced by a more rapid increase in the quantity of money than in output. The behavior of the quantity of money is the senior partner; of output, the junior partner. Many phenomena can produce temporary fluctuations in the rate of inflation, but they can have lasting effects only insofar as they affect the rate of monetary growth.

WHY THE EXCESSIVE MONETARY GROWTH?

The proposition that inflation is a monetary phenomenon is important, yet it is only the beginning of an answer to the causes of and cures for inflation. It is important because it guides the search for basic causes and limits possible cures. But it is only the beginning of an answer because the deeper question is why excessive monetary growth occurs.

Whatever was true for tobacco money or money linked to silver and gold, with today's paper money, excessive monetary growth, and hence inflation, is produced by governments.

In the United States the accelerated monetary growth during the past fifteen years or so has occurred for three related reasons: first, the rapid growth in government spending; second, the government's full employment policy; third, a mistaken policy pursued by the Federal Reserve System.

Higher government spending will not lead to more rapid monetary growth and inflation *if* additional spending is financed either by taxes or by borrowing from the public. In that case, government has more to spend, the public has less. Higher government spending is matched by lower private spending for consumption and investment. However, taxing and borrowing from the public are politically unattractive ways to finance additional government spending. Many of us welcome the additional government spending; few of us welcome additional taxes. Government borrowing from the public diverts funds from private uses by raising interest rates, making it both more expensive and more difficult for individuals to get mortgages on new homes and for businesses to borrow money.

The only other way to finance higher government spending is

by increasing the quantity of money. As we noted in Chapter 3, the U.S. government can do that by having the U.S. Treasury—one branch of the government—sell bonds to the Federal Reserve System—another branch of the government. The Federal Reserve pays for the bonds either with freshly printed Federal Reserve Notes or by entering a deposit on its books to the credit of the U.S. Treasury. The Treasury can then pay its bills with either the cash or a check drawn on its account at the Fed. When the additional high-powered money is deposited in commercial banks by its initial recipients, it serves as reserves for them and as the basis for a much larger addition to the quantity of money.

Financing government spending by increasing the quantity of money is often extremely attractive to both the President and members of Congress. It enables them to increase government spending, providing goodies for their constituents, without having to vote for taxes to pay for them, and without having to borrow from the public.

A second source of higher monetary growth in the United States in recent years has been the attempt to produce full employment. The objective, as for so many government programs, is admirable, but the results have not been. "Full employment" is a much more complex and ambiguous concept than it appears to be on the surface. In a dynamic world, in which new products emerge and old ones disappear, demand shifts from one product to another, innovation alters methods of production, and so on without end, it is desirable to have a good deal of labor mobility. People change from one job to another and often are idle for a time in between. Some people leave a job they do not like before they have found another. Young people entering the labor force take time to find jobs and experiment with different kinds of jobs. In addition, obstacles to the free operation of the labor market—trade union restrictions, minimum wages, and the like—increase the difficulty of matching worker and job. Under these circumstances, what average number of persons employed corresponds to full employment?

As with spending and taxes, there is here, too, an asymmetry. Measures that can be represented as adding to employment are politically attractive. Measures that can be represented as adding to unemployment are politically unattractive. The result is to impart a bias to government policy in the direction of adopting unduly ambitious targets of full employment.

The relation to inflation is twofold. First, government spending can be represented as adding to employment, government taxes as adding to unemployment by reducing private spending. Hence, the full employment policy reinforces the tendency for government to increase spending and lower taxes, and to finance any resulting deficit by increasing the quantity of money rather than by taxes or borrowing from the public. Second, the Federal Reserve System can increase the quantity of money in ways other than financing government spending. It can do so by buying outstanding government bonds, paying for them with newly created high-powered money. That enables the banks to make a larger volume of private loans, which can also be represented as adding to employment. Under pressure to promote full employment, the Fed's monetary policy has had the same inflationary bias as the government's fiscal policy.

These policies have not succeeded in producing full employment but they have produced inflation. As Prime Minister James Callaghan put it in a courageous talk to a British Labour party conference in September 1976: "We used to think that you could just spend your way out of a recession and increase employment by cutting taxes and boosting government spending. I tell you, in all candor, that that option no longer exists; and that insofar as it ever did exist, it only worked by injecting bigger doses of inflation into the economy followed by higher levels of unemployment as the next step. That is the history of the past twenty years."

The third source of higher monetary growth in the United States in recent years has been a mistaken policy by the Federal Reserve System. Not only has the Fed's policy had an inflationary bias because of pressures to promote full employment, but that bias has been exacerbated by its attempt to pursue two incompatible objectives. The Fed has the power to control the quantity of money and it gives lip service to that objective. But like Demetrius in Shakespeare's *A Midsummer Night's Dream*, who shuns Helena, who is in love with him, to pursue Hermia, who loves another, the Fed has given its heart not to controlling the quantity of money but to controlling interest rates, something that it does not have the power to do. The result has been failure on both fronts: wide swings in both money and interest rates. These swings, too, have had an inflationary bias. With memories of its disastrous mistake from 1929 to 1933, the Fed has been much prompter in correcting a swing toward a low rate of monetary growth than in correcting a swing toward a high rate of monetary growth.

The end result of higher government spending, the full employment policy, and the Fed's obsession with interest rates has been a roller coaster along a rising path. Inflation has risen and then fallen. Each rise has carried inflation to a higher level than the preceding peak. Each fall has left inflation above its preceding trough. All the time, government spending has been rising as a fraction of income; government tax receipts, too, have been rising as a fraction of income, but not quite as fast as spending, so the deficit, too, has been rising as a fraction of income.

These developments are not unique to the United States or to recent decades. Since time immemorial, sovereigns—whether kings, emperors, or parliaments—have been tempted to resort to increasing the quantity of money to acquire resources to wage wars, construct monuments, or for other purposes. They have often succumbed to the temptation. Whenever they have, inflation followed close behind.

Nearly two thousand years ago the Roman Emperor Diocletian inflated by "debasing" the coinage—that is, replacing silver coins by look-alikes that had less and less silver and more and more of a worthless alloy until they became "no more than base metal washed over with silver." [12] Modern governments do so by printing paper money and making entries on books—but the ancient method has not entirely disappeared. The once full-bodied silver coins of the United States are now copper coins washed over, not even with silver, but with nickel. And a small-size Susan B. Anthony dollar coin has been introduced to replace what was once a full-bodied silver coin.

GOVERNMENT REVENUE FROM INFLATION

Financing government spending by increasing the quantity of money looks like magic, like getting something for nothing. To take a simple example, government builds a road, paying for the expenses incurred with newly printed Federal Reserve Notes. It looks as if everybody is better off. The workers who build the road get their pay and can buy food, clothing, and housing with it. Nobody has paid higher taxes. Yet there is now a road where there was none before. Who has paid for it?

The answer is that all holders of money have paid for the road. The extra money raises prices when it is used to induce the workers to build the road instead of engage in some other productive activity. Those higher prices are maintained as the extra money circulates in the spending stream from the workers to the sellers of what they buy, from those sellers to others, and so on. The higher prices mean that the money people previously held will now buy less than it would have before. In order to have on hand an amount of money that can buy as much as before, they will have to refrain from spending all of their income and use part of it to add to their money balances.

The extra money printed is equivalent to a tax on money balances. If the extra money raises prices by 1 percent, then every holder of money has in effect paid a tax equal to 1 percent of his money holdings. The extra pieces of paper he now must hold (or book entries he must make) in order to have the same purchasing power in the form of money as before are indistinguishable from the other pieces of paper in his pocket or safe deposit box (or from book entries), but they are in effect receipts for taxes paid.

The physical counterpart to these taxes is the goods and services that could have been produced by the resources that built the road. The people who spent less than their income in order to maintain the purchasing power of their money balances have given up these goods and services in order that the government could get the resources to build the road.

You can see why John Maynard Keynes, in discussing the inflations after World War I, wrote: "There is no subtler, no surer means of overturning the existing basis of society than to debauch the currency. The process engages all the hidden forces of economic law on the side of destruction, and does it in a manner which not one man in a million is able to diagnose." [13]

The additional currency printed and the additional deposits

entered on the books of the Federal Reserve Bank correspond to only part of the revenue that government gets from inflation.

Inflation also yields revenue indirectly by automatically raising effective tax rates. As people's dollar incomes go up with inflation, the income is pushed into higher brackets and taxed at a higher rate. Corporate income is artificially inflated by inadequate allowance for depreciation and other costs. On the average, if income rises by 10 percent simply to match a 10 percent inflation, federal tax revenue tends to go up by more than 15 percent—so the taxpayer has to run faster and faster to stay in the same place. That process has enabled the President, Congress, state governors and legislatures to pose as tax cutters when all they have done is to keep taxes from going up as much as they otherwise would have gone up. Each year, there is talk of "cutting taxes." Yet there has been no reduction in taxes. On the contrary, taxes correctly measured have gone up—at the federal level from 22 percent of national income in 1964 to 25 percent in 1978; at the state and local level from 11 percent in 1964 to 15 percent in 1978.

Still a third way inflation yields revenue to the government is by paying off—or repudiating, if you will—part of the government's debt. Government borrows in dollars and pays back in dollars. But thanks to inflation, the dollars it pays back can buy less than the dollars it borrowed. That would not be a net gain to the government if in the interim it had paid a high enough interest rate on the debt to compensate the lender for inflation. But for the most part it did not. Savings bonds are the clearest example. Suppose you had bought a savings bond in December 1968, had held it to December 1978, and then cashed it in. You would have paid $37.50 in 1968 for a ten-year bond with a face value of $50 and you would have received $64.74 when you cashed it in 1978 (because the government raised the interest rate in the interim to make some allowance for inflation). By 1978 it took $70 to buy as much as $37.50 would have bought in 1968. Yet not only would you have gotten back only $64.74, you would have had to pay income tax on the $27.24 difference between what you received and what you paid. You would have ended up paying for the dubious privilege of lending to your government.

Paying off the debt by inflation has meant that although the federal government has run large deficits year after year and its debt in terms of dóllars has gone up, the debt has gone up far less in terms of purchasing power and has actually fallen as a percentage of the national income. In the decade from 1968 through 1978, the federal government had a cumulative deficit of more than $260 billion, yet the debt amounted to 30 percent of national income in 1968, to 28 percent in 1978.

THE CURE FOR INFLATION

The cure for inflation is simple to state but hard to implement. Just as an excessive increase in the quantity of money is the one and only important cause of inflation, so a reduction in the rate of monetary growth is the one and only cure for inflation. The problem is not one of knowing what to do. That is easy enough. Government must increase the quantity of money less rapidly. The problem is to have the political will to take the measures necessary. Once the inflationary disease is in an advanced state, the cure takes a long time and has painful side effects.

Two medical analogies suggest the problem. One is about a young man who had Buerger's disease, a disease that interrupts the blood supply and can lead to gangrene. The young man was losing fingers and toes. The cure was simple to state: stop smoking. The young man did not have the will to do so; his addiction to tobacco was simply too great. His disease was in one sense curable, in another not.

A more instructive analogy is between inflation and alcoholism. When the alcoholic starts drinking, the good effects come first; the bad effects only come the next morning when he wakes up with a hangover—and often cannot resist easing the hangover by taking "the hair of the dog that bit him."

The parallel with inflation is exact. When a country starts on an inflationary episode, the initial effects seem good. The increased quantity of money enables whoever has access to it—nowadays, primarily governments—to spend more without anybody else having to spend less. Jobs become more plentiful, business is brisk, almost everybody is happy—at first. Those are the good effects. But then the increased spending starts to raise prices; workers find that their wages, even if higher in dollars, will buy less; businessmen find that their costs have risen, so that the extra sales are not as profitable as they anticipated, unless they can raise their prices even faster. The bad effects start to emerge: higher prices, less buoyant demand, inflation combined with stagnation. As with the alcoholic, the temptation is to increase the quantity of money still faster, which produces the roller coaster we have been on. In both cases, it takes a larger and larger amount—of alcohol or money—to give the alcoholic or the economy the same "kick."

The parallel between alcoholism and inflation carries over to the cure. The cure for alcoholism is simple to state: stop drinking. It is hard to take because, this time, the bad effects come first, the good effects come later. The alcoholic who goes on the wagon suffers severe withdrawal pains before he emerges in the happy land of no longer having an almost irresistible desire for another drink. So also with inflation. The initial side effects of a slower rate of monetary growth are painful: lower economic growth, temporarily high unemployment, without, for a time, much reduction of inflation. The benefits appear only after one or two years or so, in the form of lower inflation, a healthier economy, the potential for rapid noninflationary growth.

Painful side effects are one reason why it is difficult for an alcoholic or an inflationary nation to end its addiction. But there is another reason, which, at least in the earlier stage of the disease, may be even more important: the lack of a real desire to end the addiction. The drinker enjoys his liquor; he finds it hard to accept that he really is an alcoholic; he is not sure he wants to take the cure. The inflationary nation is in the same position. It is tempting to believe that inflation is a temporary and mild matter produced by unusual and extraneous circumstances, and that it will go away of its own accord—something that never happens.

Moreover, many of us enjoy inflation. We would naturally like to see the prices of the things we *buy* go down, or at least stop going up. But we are more than happy to see the prices of the things we *sell* go up—whether goods we produce, our labor services, or houses or other items we own. Farmers complain about inflation but congregate in Washington to lobby for higher prices for their products. Most of the rest of us do the same in one way or another.

One reason inflation is so destructive is because some people benefit greatly while other people suffer; society is divided into winners and losers. The winners regard the good things that happen to them as the natural result of their own foresight, prudence, and initiative. They regard the bad things, the rise in the prices of the things they buy, as produced by forces outside their control. Almost everyone will say that he is against inflation; what he generally means is that he is against the bad things that have happened to him.

To take a specific example, almost every person who has owned a home during the past two decades has benefited from inflation. The value of his home has risen sharply. If he had a mortgage, the interest rate was generally below the rate of inflation. As a result the payments called "interest," as well as those called "principal," have in effect been paying off the mortgage. To take a simple example, suppose both the interest rate and inflation rate were 7 percent in one year. If you had a $10,000 mortgage on which you paid only interest, a year later the mortgage would correspond to the same buying power as $9,300 would have a year earlier. In real terms you would owe $700 less—just the amount you paid as interest. In real terms you would have paid nothing for the use of the $10,000. (Indeed, because the interest is deductible in computing your income tax, you would actually benefit. You would have been paid for borrowing.) The way this effect becomes apparent to the homeowner is that his equity in the house goes up rapidly. The counterpart is a loss to the small savers who provided the funds that enabled savings and loan associations, mutual savings banks, and other institutions to finance mortgage loans. The small savers had no good alternative because government limits narrowly the maximum interest rate that such institutions can pay to their depositors—supposedly to protect the depositors.

Just as high government spending is one reason for excessive monetary growth, so lower government spending is one element that can contribute to reducing monetary growth. Here, too, we tend to be schizophrenic. We would all like to see government spending go down, provided it is not spending that benefits us. We would all like to see deficits reduced, provided it is through taxes imposed on others.

As inflation accelerates, however, sooner or later it does so much damage to the fabric of society, creates so much injustice and suffering, that a real public will develops to do something about inflation. The level of inflation at which that occurs depends critically on the country in question and its history. In Germany it came at a low level of inflation because of Germany's terrible experiences after World War I and II; it came at a much higher level of inflation in the United Kingdom and Japan; it has not yet come in the United States.

SIDE EFFECTS OF A CURE

We read over and over again that higher unemployment and slow growth are cures for inflation, that the alternatives we must face are more inflation *or* higher unemployment, that the powers that be are reconciled to, or are positively promoting, slower growth and higher unemployment in order to cure inflation. Yet over the past several decades, the growth of the U.S. economy has slowed, the average level of unemployment has risen, and at the same time, the rate of inflation has moved higher and higher. We have had both more inflation and more unemployment. Other countries have had the same experience. How come?

The answer is that slow growth and high unemployment are not *cures* for inflation. They are *side effects* of a successful cure. Many policies that impede economic growth and add to unemployment may, at the same time, increase the rate of inflation. That has been true of some of the policies we have adopted— sporadic price and wage control, increasing government intervention into business, all accompanied by higher and higher government spending, and a rapid increase in the quantity of money.

Another medical example will perhaps make clear the difference between a *cure* and a *side effect*. You have acute appendicitis. Your physician recommends an appendectomy but warns you that after the operation you will be confined to bed for an interval. You refuse the operation but take to your bed for the indicated interval as a less painful *cure*. Silly, yes, but faithful in every detail to the confusion between unemployment as a side effect and as a cure.

The side effects of a cure for inflation are painful so it is important to understand why they occur and to seek means to mitigate them. The basic reason why the side effects occur has already been pointed out in Chapter 1. They occur because variable rates of monetary growth introduce static into the information transmitted by the price system, static that is translated into inappropriate responses by the economic actors, which it takes time to overcome.

Consider, first, what happens when inflationary monetary growth starts. The higher spending financed by the newly created money is no different to the seller of goods or labor or other services from any other spending. The seller of pencils, for example, finds that he can sell more pencils at the former price. He does so initially without changing his price. He orders more pencils from the wholesaler, the wholesaler from the manufacturer, and so on down the line. *If* the demand for pencils had increased at the expense of some other segment of demand, say at the expense of the demand for ball-point pens, rather than as a result of inflationary monetary growth, the increased flow of orders down the pencil channel would be accompanied by a decreased flow down the ball-point pen channel. Pencils and later the materials used to make them would tend to rise in price; pens and the materials used to make them would tend to fall in price; but there would be no reason for prices *on the average* to change.

The situation is wholly different when the increased demand for pencils has its origin in newly created money. The demand for pencils and pens and most other things can then go up simultaneously. There is more spending (in dollars) in total. However, the seller of pencils does not know this. He proceeds as before, initially holding the price at which he sells constant, content to sell more until, as he believes, he will be able to restock. But now the increased flow of orders down the pencil channel is accompanied by an increased flow down the pen channel, and down many other channels. As the increased flow of orders generates a greater demand for labor and materials to produce more, the initial reaction of workers and producers of materials will be like that of the retailers—to work longer and produce more and also charge more in the belief that the demand for what they have been providing has gone up. But this time there is no offset, there are no declines in demand roughly matching the increases in demand, no declines in prices matching the increases. Of course, this will not at first be obvious. In a dynamic world demands are always shifting, some prices going up, some going down. The general signal of increasing demand will be confused with the specific signals reflecting changes in relative demands. That is why the initial side effect of faster monetary growth is an appearance of prosperity and greater employment. But sooner or later the signal will get through.

As it does, workers, manufacturers, retailers will discover that they have been fooled. They reacted to higher demand for the small number of things they sell in the mistaken belief that the higher demand was special to them and hence would not much affect the prices of the many things they buy. When they discover their mistake, they raise wages and prices still higher—not only to respond to higher demand but also to allow for the rises in the prices of the things they buy. We are off on a price-wage spiral that is itself an effect of inflation, not a cause. If monetary growth does not speed up further, the initial stimulus to employment and output will be replaced by the opposite; both will tend to go down in response to the higher wages and prices. A hangover will succeed the initial euphoria.

It takes time for these reactions to occur. On the average over the past century and more in the United States, the United Kingdom, and some other Western countries, roughly six to nine months have elapsed before increased monetary growth has worked its way through the economy and produced increased economic growth and employment. Another twelve to eighteen months have elapsed before the increased monetary growth has affected the price level appreciably and inflation has occurred or speeded up. The time delays have been this long for these countries because, wartime aside, they were long spared widely varying rates of monetary growth and inflation. On the eve of World War II wholesale prices in the United Kingdom averaged roughly the same as two hundred years earlier, and in the United States, as one hundred years earlier. The post–World War II inflation is a new phenomenon in these countries. They have experienced many ups and downs but not a long movement in the same direction.

Many countries in South America have had a less happy heritage. They experience much shorter time delays—amounting at most to a few months. If the United States does not cure its recent propensity to indulge in widely varying rates of inflation, the time delays will shorten here as well.

The sequence of events that follows a slowing of monetary growth is the same as that just outlined except in the opposite direction. The initial reduction in spending is interpreted as a reduction in demand for specific products, which after an interval leads to a reduction in output and employment. After another interval inflation slows, which in turn is accompanied by an expansion in employment and output. The alcoholic is through his worst withdrawal pains and on the road to contented abstinence.

All of these adjustments are set in motion by *changes* in the rates of monetary growth and inflation. If monetary growth were high and steady, so that, let us say, prices tended to rise year after year by 10 percent, the economy could adjust to it. Everybody would come to anticipate a 10 percent inflation; wages would rise by 10 percent a year more than they otherwise would; interest rates would be 10 percentage points higher than otherwise—in order to compensate the lender for inflation; tax rates would be adjusted for inflation, and so on and on.

Such an inflation would do no great harm, but neither would it serve any function. It would simply introduce unnecessary complexities in arrangements. More important, such a situation, if it ever developed, would probably not be stable. If it were politically profitable and feasible to generate a 10 percent inflation, the temptation would be great, when and if inflation ever settled there, to make the inflation 11 or 12 or 15 percent. Zero inflation is a politically feasible objective; a 10 percent inflation is not. That is the verdict of experience.

MITIGATING THE SIDE EFFECTS

We know no example in history in which an inflation has been ended without an interim period of slow economic growth and higher than usual unemployment. That is the basis in experience for our judgment that there is no way to avoid side effects of a cure for inflation.

However, it is possible to mitigate those side effects, to make them milder.

The most important device for mitigating the side effects is to slow inflation *gradually but steadily* by a policy announced in advance and adhered to so it becomes credible.

The reason for gradualness and advance announcement is to give people time to readjust their arrangements—and to induce them to do so. Many people have entered into long-term contracts—for employment, to lend or borrow money, to engage in production or construction—on the basis of *anticipations* about the likely rate of inflation. These long-term contracts make it difficult to reduce inflation rapidly and mean that trying to do so will impose heavy costs on many people. Given time, these contracts will be completed or renewed or renegotiated, and can then be adjusted to the new situation.

One other device has proved effective in mitigating the adverse side effects of curing inflation—including an automatic adjustment for inflation in longer-term contracts, what are known as escalator clauses. The most common example is the cost-of-living adjustment clause that is included in many wage contracts. Such a contract specifies that the hourly wage shall increase by, say, 2 percent plus the rate of inflation or plus a fraction of the rate of inflation. In that way, if inflation is low, the wage increase in dollars is low; if inflation is high, the wage increase in dollars is high; but in either case the wage has the same purchasing power.

Another example is for contracts for the rental of property. Instead of being stated as a fixed number of dollars, the rental contract may specify that the rent shall be adjusted from year to year by the rate of inflation. Rental contracts for retail stores often specify the rent as a percentage of the gross receipts of the store. Such contracts have no explicit escalator clause but implicitly they do, since the store's receipts will tend to rise with inflation.

Still another example is for a loan. A loan is typically for a fixed dollar sum for a fixed period at a fixed annual rate of interest, say, $1,000 for one year at 10 percent. An alternative is to specify the rate of interest not at 10 percent but, say, 2 percent plus the rate of inflation, so that if inflation turns out to be 5 percent, the interest rate will be 7 percent; if inflation turns out to be 10 percent, the interest rate will be 12 percent. An alternative that is roughly equivalent is to specify the amount to be repaid not as a fixed number of dollars but as a number of dollars adjusted for inflation. In our simple example the borrower would owe $1,000 increased by the rate of inflation plus interest at 2 percent. If inflation turned out to be 5 percent, he would owe $1,050; if 10 percent, $1,100; in both cases plus interest at 2 percent.

Except for wage contracts, escalator clauses have not been common in the United States. However, they are spreading, especially in the form of variable interest mortgages. And they have been common in just about all countries that have experienced both high and variable rates of inflation over any extensive period.

Such escalator clauses reduce the time delay between slowing down monetary growth and the subsequent adjustment of wages and prices. In that way they shorten the transition period and reduce the interim side effects. However, useful though they are, escalator clauses are far from a panacea. It is impossible to escalate *all* contracts (consider, for example, paper money), and costly to escalate many. A major advantage of using money is precisely the ability to carry on transactions cheaply and efficiently, and universal escalator clauses reduce this advantage. Far better to have no inflation and no escalator clauses. That is why we advocate resort to escalator clauses in the private economy only as a device for easing the side effects of curing inflation, not as a permanent measure.

Escalator clauses are highly desirable as a permanent measure in the federal government sector. Social Security and other retirement benefits, salaries of federal employees, including the salaries of members of Congress, and many other items of government spending are now automatically adjusted for inflation. However, there are two glaring and inexcusable gaps: income taxes and government borrowing. Adjusting the personal and corporate tax structure for inflation—so that a 10 percent price rise would raise taxes in dollars by 10 percent, not, as it does now, by something over 15 percent on the average—would eliminate the imposition of higher taxes without their having been voted. It would end this taxation without representation. By so doing, it would also reduce the incentive for the government to inflate, since the revenue from inflation would be reduced.

The case for inflation-proofing government borrowing is equally strong. The U.S. government has itself produced the inflation that has made the purchase of long-term government bonds such a poor investment in recent years. Fairness and honesty toward citizens on the part of their government require introducing escalator clauses into long-term government borrowing.

Price and wage controls are sometimes proposed as a cure for inflation. Recently, as it has become clear that controls are not a cure, they have been urged as a device for mitigating the side effects of a cure. It is claimed that they will serve this function by persuading the public that the government is serious in attacking inflation. That, in turn, is expected to lower the anticipations of future inflation that are built into the terms of long-term contracts.

Price and wage controls are counterproductive for this purpose. They distort the price structure, which reduces the efficiency with which the system works. The resulting lower output adds to the adverse side effects of a cure for inflation rather than reducing them. Price and wage controls waste labor, both because of the distortions in the price structure and because of the immense amount of labor that goes into constructing, enforcing, and evading the price and wage controls. These effects are the same whether controls are compulsory or are labeled "voluntary."

In practice, price and wage controls have almost always been used as a substitute for monetary and fiscal restraint, rather than as a complement to them. This experience has led participants in the market to regard the imposition of price and wage controls as a signal that inflation is heading up, not down. It has therefore led them to raise their inflation expectations rather than to lower them.

Price and wage controls often seem effective for a brief period after they are imposed. Quoted prices, the prices that enter into index numbers, are kept down because there are indirect ways of raising prices and wages—lowering the quality of items produced, eliminating services, promoting workers, and so on. But then, as the easy ways of avoiding the controls are exhausted, distortions accumulate, the pressures suppressed by the controls reach the boiling point, the adverse effects get worse and worse, and the whole program breaks down. The end result is more inflation, not less. In light of the experience of forty centuries, only the short time perspective of politicians and voters can explain the repeated resort to price and wage controls.[14]

A CASE STUDY

Japan's recent experience provides an almost textbook illustration of how to cure inflation. As Figure 6 shows, the quantity of money in Japan began growing at higher and higher rates in 1971, and by mid-1973, it was growing more than 25 percent a year.[15]

Inflation did not respond until about two years later, in early 1973. The subsequent dramatic rise in inflation produced a funda-

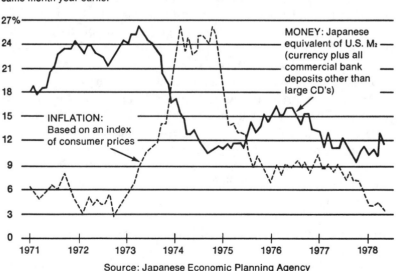

Figure 6. INFLATION FOLLOWS MONEY:
THE CASE OF JAPAN

Percent increase from
same month year earlier

MONEY: Japanese
equivalent of U.S. M₂
(currency plus all
commercial bank
deposits other than
large CD's)

INFLATION:
Based on an index
of consumer prices

Source: Japanese Economic Planning Agency

mental change in monetary policy. Emphasis shifted from the external value of the yen—the exchange rate—to its internal value —inflation. Monetary growth was reduced sharply, from more than 25 percent a year to between 10 and 15 percent. It was kept there, with minor exceptions, for five years. (Because of Japan's high rate of economic growth, monetary growth in this range would produce roughly stable prices. The comparable rate for the United States is 3 to 5 percent.)

About eighteen months after monetary growth started declining, inflation followed suit, but it took two and a half years before inflation fell below double digits. Inflation then held roughly constant for about two years—despite a mild upturn in monetary growth. Inflation then started moving rapidly toward zero in response to a new decline in monetary growth.

The numbers on inflation in the chart are for consumer prices. Wholesale prices did even better. They actually declined after mid-1977. The postwar shift of workers in Japan from low-productivity sectors to high productivity sectors, such as automobiles and electronics, has meant that prices of services have risen sharply relative to prices of commodities. As a result, consumer prices have risen relative to wholesale prices.

Japan experienced lower growth and higher unemployment after it slowed monetary growth, particularly during 1974 before inflation started to respond appreciably to the slower monetary growth. The low point was reached at the end of 1974. Output then began recovering and grew thereafter—more modestly than in the boom years of the 1960s but at a highly respectable rate nonetheless: more than 5 percent per year.

Price and wage controls were not imposed at any time during the tapering down of inflation. And the tapering down occurred at the same time that Japan was adjusting to higher prices for crude oil.

CONCLUSIONS

Five simple truths embody most of what we know about inflation:

1. Inflation is a monetary phenomenon arising from a more

rapid increase in the quantity of money than in output (though, of course, the reasons for the increase in money may be various).

2. In today's world government determines—or can determine —the quantity of money.

3. There is only one cure for inflation: a slower rate of increase in the quantity of money.

4. It takes time—measured in years, not months—for inflation to develop; it takes time for inflation to be cured.

5. Unpleasant side effects of the cure are unavoidable.

The United States has embarked on rising monetary growth four times during the past twenty years. Each time the higher monetary growth has been followed first by economic expansion, later by inflation. Each time the authorities have slowed monetary growth in order to stem inflation. Lower monetary growth has been followed by an inflationary recession. Later still, inflation has declined and the economy has improved. So far the sequence is identical with Japan's experience from 1971 to 1975. Unfortunately, the crucial difference is that we have not displayed the patience Japan did by continuing monetary restraint long enough. Instead, we have overreacted to the recession by accelerating monetary growth, setting off on another round of inflation, and condemning ourselves to higher inflation plus higher unemployment.

We have been misled by a false dichotomy: inflation or unemployment. That option is an illusion. The real option is only whether we have higher unemployment as a result of higher inflation or as a temporary side effect of curing inflation.

CHAPTER 10

The Tide
Is Turning

The failure of Western governments to achieve their proclaimed objectives has produced a widespread reaction against big government. In Britain the reaction swept Margaret Thatcher to power in 1979 on a platform pledging her Conservative government to reverse the socialist policies that had been followed by both Labour and earlier Conservative governments ever since the end of World War II. In Sweden in 1976, the reaction led to the defeat of the Social Democratic party after more than four decades of uninterrupted rule. In France the reaction led to a dramatic change in policy designed to eliminate government control of prices and wages and sharply reduce other forms of government intervention. In the United States the reaction has been manifested most dramatically in the tax revolt that has swept the nation, symbolized by the passage of Proposition 13 in California, and realized in a number of states in constitutional amendments limiting state taxes.

The reaction may prove short-lived and be followed, after a brief interval, by a resumption of the trend toward ever bigger government. The widespread enthusiasm for reducing government taxes and other impositions is not matched by a comparable enthusiasm for eliminating government programs—except programs that benefit other people. The reaction against big government has been sparked by rampant inflation, which governments can control if they find it politically profitable to do so. If they do, the reaction might be muted or disappear.

We believe that the reaction is more than a response to transitory inflation. On the contrary, the inflation itself is partly a response to the reaction. As it has become politically less attractive to vote higher taxes to pay for higher spending, legislators have resorted to financing spending through inflation, a hidden tax that can be imposed without having been voted, taxation without representation. That is no more popular in the twentieth century than it was in the eighteenth.

In addition, the contrast between the ostensible objectives of government programs and their actual results—a contrast that has been a persistent theme of earlier chapters—is so pervasive, so widespread, that even many of the strongest supporters of big government have had to acknowledge government failure—though their solution almost always turns out to be still bigger government.

A tide of opinion, once it flows strongly, tends to sweep over all obstacles, all contrary views. Equally, when it has crested and a contrary tide sets in, that too tends to flow strongly.

The tide of opinion toward economic freedom and limited government that Adam Smith and Thomas Jefferson did so much to promote flowed strongly until late in the nineteenth century. Then the tide of opinion turned—in part because the very success of economic freedom and limited government in producing economic growth and improving the well-being of the bulk of the population rendered the evils that remained (and of course there were many) all the more prominent and evoked a widespread desire to do something about them. The tide toward Fabian socialism and New Deal liberalism in turn flowed strongly, fostering a change in the direction of British policy early in the twentieth century, and in U.S. policy after the Great Depression.

That trend has now lasted three-quarters of a century in Britain, half a century in the United States. It, too, is cresting. Its intellectual basis has been eroded as experience has repeatedly contradicted expectations. Its supporters are on the defensive. They have no solutions to offer to present-day evils except more of the same. They can no longer arouse enthusiasm among the young who now find the ideas of Adam Smith or Karl Marx far more exciting than Fabian socialism or New Deal liberalism.

Though the tide toward Fabian socialism and New Deal liberalism has crested, there is as yet no clear evidence whether the tide that succeeds it will be toward greater freedom and limited government in the spirit of Smith and Jefferson or toward an omnipotent monolithic government in the spirit of Marx and Mao. That vital matter has not yet been determined—either for the intellectual climate of opinion or for actual policy. To judge from the past, it will be determined for opinion first and policy will then follow suit.

IMPORTANCE OF INTELLECTUAL CLIMATE OF OPINION

The example of India and Japan, discussed in Chapter 2, exemplifies the importance of the intellectual climate of opinion, which determines the unthinking preconceptions of most people and their leaders, their conditioned reflexes to one course of action or another.

The Meiji leaders who took charge of Japan in 1867 were dedicated primarily to strengthening the power and glory of their country. They attached no special value to individual freedom or political liberty. They believed in aristocracy and political control by an elite. Yet they adopted a liberal economic policy that led to the widening of opportunities for the masses and, during the early decades, greater personal liberty. The men who took charge in India, on the other hand, were ardently devoted to political freedom, personal liberty, and democracy. Their aim was not only national power but also improvement in the economic conditions of the masses. Yet they adopted a collectivist economic policy that hamstrings their people with restrictions and continues to undermine the large measure of individual freedom and political liberty encouraged by the British.

The difference in policies reflects faithfully the different intellectual climates of the two eras. In the mid-nineteenth century it was taken for granted that a modern economy should be organized through free trade and private enterprise. It probably never occurred to the Japanese leaders to follow any other course. In the mid-twentieth century, it was taken for granted that a modern economy should be organized through centralized control and five-year plans. It probably never occurred to the Indian leaders to follow any other course. It is an interesting sidelight that both views came from Great Britain. The Japanese adopted the policies of Adam Smith. The Indians adopted the policies of Harold Laski.

Our own history is equally strong evidence of the importance of the climate of opinion. It shaped the work of the remarkable group of men who gathered in Independence Hall in Philadelphia in 1787 to write a constitution for the new nation they had helped to create. They were steeped in history and were greatly influenced by the current of opinion in Britain—the same current that was later to affect Japanese policy. They regarded concentration of power, especially in the hands of government, as the great danger to freedom. They drafted the Constitution with that in mind. It was a document intended to limit government power, to keep power decentralized, to reserve to individuals control over their own lives. This thrust is even clearer in the Bill of Rights, the first ten amendments to the Constitution, than in the basic text: "Congress shall make no law respecting an establishment of religion, or prohibiting the free exercise thereof; or abridging the freedom of speech, or of the press"; "the right of the people to keep and bear arms shall not be infringed"; "the enumeration in the Constitution, of certain rights, shall not be construed to deny or disparage others retained by the people"; "the powers not delegated to the United States by the Constitution, nor prohibited by it to the States, are reserved to the States, respectively, or to the people" (from Amendments I, II, IX, and X).

Late in the nineteenth century and on into the early decades of the twentieth, the intellectual climate of opinion in the United States—largely under the influence of the same views from Britain that later affected Indian policy—started to change. It moved away from a belief in individual responsibility and reliance on the market toward a belief in social responsibility and reliance on the government. By the 1920s a strong minority, if not an actual majority, of college and university professors actively concerned with public affairs held socialist views. The *New Republic* and the *Nation* were the leading intellectual journals of opinion. The Socialist party of the United States, led by Norman Thomas, had broader roots, but much of its strength was in colleges and universities.

In our opinion the Socialist party was the most influential political party in the United States in the first decades of the twentieth century. Because it had no hope of electoral success on a national level (it did elect a few local officials, notably in Milwaukee, Wisconsin), it could afford to be a party of principle. The Democrats and Republicans could not. They had to be parties of expediency and compromise, in order to hold together widely disparate factions and interests. They had to avoid "extremism," keep to the middle ground. They were not exactly Tweedledum and Tweedledee—but close to it. Nonetheless, in the course of time both major parties adopted the position of the Socialist party. The Socialist party never received more than 6 percent of the popular vote for President (in 1912 for Eugene Debs). It got less than 1 percent in 1928 and only 2 percent in 1932 (for Norman Thomas). Yet almost every economic plank in its 1928 presidential platform has by now been enacted into law. The relevant planks are reproduced in Appendix A.

Once the change in the climate of opinion had spread to a wider public, as it did after the Great Depression, the Constitution shaped by a very different climate of opinion proved at most a source of delay to the growth of government power, not an obstacle.

In Mr. Dooley's words, "No matter whether th' constitution follows th' flag or not, th' supreme court follows th' iliction returns." The words of the Constitution were reinterpreted and given new meaning. What had been intended to be barriers to the extension of government power were rendered ineffective. As Raoul Berger writes in his authoritative examination of the Court's interpretation of one amendment,

> The Fourteenth Amendment is the case study par excellence of what Justice Harlan described as the Supreme Court's "exercise of the amending power," its continuing revision of the Constitution under the guise of interpretation. . . .
> The Court, it is safe to say, has flouted the will of the framers and substituted an interpretation in flat contradiction of the original design. . . .
> Such conduct impels one to conclude that the Justices are become a law unto themselves.[1]

OPINION AND POPULAR BEHAVIOR

Evidence that the tide toward Fabian socialism and New Deal liberalism has crested comes not only from the writing of intellectuals, not only from the sentiments that politicians express on the hustings, but also from the way people behave. Their behavior is no doubt influenced by opinion. In its turn, popular behavior both reinforces that opinion and plays a major role in translating it into policy.

As A. V. Dicey, with remarkable prescience, wrote more than sixty years ago, "If the progress of socialistic legislation be arrested, the check will be due, not so much to the influence of any thinker as to some patent fact which shall command public attention; such, for instance, as that increase in the weight of taxation which is apparently the usual, if not the invariable, concomitant of a socialist policy." [2] Inflation, high taxes, and the patent inefficiency, bureaucracy, and excessive regulation stemming from big government are having the effects Dicey predicted. They are leading people to take matters into their own hands, to try to find ways around government obstacles.

Pat Brennan became something of a celebrity in 1978 because she and her husband went into competition with the U.S. Post Office. They set up business in a basement in Rochester, New York, guaranteeing delivery the same day of parcels and letters in downtown Rochester at a lower cost than the Post Office charged. Soon their business was thriving.

There is no doubt that they were breaking the law. The Post Office took them to court, and they lost after a legal battle that went all the way to the Supreme Court. Local businessmen provided financial backing.

Said Pat Brennan,

> I think there's going to be a quiet revolt and perhaps we're the beginning of it. . . . You see people bucking the bureaucrats, when years ago you wouldn't dream of doing that because you'd be squashed. . . . People are deciding that their fates are their own and not up to somebody in Washington who has no interest in them whatsoever. So it's not a question of anarchy, but it's a question of people rethinking the power of the bureaucrats and rejecting it. . . .
>
> The question of freedom comes up in any kind of a business— whether you have the right to pursue it and the right to decide what you're going to do. There is also the question of the freedom of the consumers to utilize a service that they find is inexpensive and far superior, and according to the federal government and the body of laws called the Private Express Statutes, I don't have the freedom to start a business and the consumer does not have the freedom to use it—which seems very strange in a country like this that the entire context of the country is based on freedom and free enterprise.

Pat Brennan is expressing a natural human response to the attempt by other people to control her life when she thinks it's none of their business. The first reaction is resentment; the second is to attempt to get around obstacles by legal means; finally, there comes a decline in respect for law in general. This final consequence is deplorable but inevitable.

A striking example is what has happened in Great Britain in reaction to confiscatory taxes. Says a British authority, Graham Turner:

> I think that it's perfectly fair to say that we have become in the course of the last ten or fifteen years a nation of fiddlers.
>
> How do they do it? They do it in a colossal variety of ways. Let's take it right at the lowest level. Take a small grocer in a country area, . . . how does he make money? He finds out that by buying through regular wholesalers he's always got to use invoices, but if he goes to the Cash and Carry and buys his goods from there, . . . the profit margin on those goods can be untaxed because the tax inspectors simply don't know that he's had those goods. That's the way he does it.
>
> Then if you take it at the top end—if you take a company director —well, there are all kinds of ways that they can do it. They buy their food through the company, they have their holidays on the company, they put their wives as company directors even though they never visit the factory. They build their houses on the company by a very simple device of building a factory at the same time as a house.
>
> It goes absolutely right through the range, from the ordinary working-class person doing quite menial jobs, right to the top end—businessmen, senior politicians, members of the Cabinet, members of the Shadow Cabinet—they all do it.
>
> I think almost everybody now feels that the tax system is basically unfair, and everybody who can, tries to find a way round that tax system. Now once there's a consensus that a tax system is unfair, the country in effect becomes a kind of conspiracy—and everybody helps each other to fiddle.
>
> You've no difficulty fiddling in this country because other people actually want to help you. Now fifteen years ago that would have been quite different. People would have said, hey, this is not quite as it should be.

Or consider this, from an article in the *Wall Street Journal* by Melvyn B. Krauss on "The Swedish Tax Revolt" (February 1, 1979, p. 18):

The Swedish revolution against the highest taxes in the West is based on individual initiative. Instead of relying on politicians, ordinary Swedes have taken matters into their own hands and simply refuse to pay. This can be done in several ways, many of them legal. . . .

One way a Swede refuses to pay taxes is by working less. . . . Swedes sailing in Stockholm's beautiful archipelago vividly illustrate the country's quiet tax revolution.

The Swedes escape tax by doing-it-themselves. . . .

Barter is another way Swedes resist high taxes. To entice a Swedish dentist off the tennis court and into his office is no easy matter. But a lawyer with a toothache has a chance. The lawyer can offer legal services in return for dental services. Bartering saves the dentist two taxes: his own income tax plus the tax on the lawyer's fees. Though barter is supposed to be a sign of a primitive economy, high Swedish taxes have made it a popular way of doing business in the welfare state, particularly in the professions. . . .

The tax revolution in Sweden is not a rich man's revolution. It is taking place at all income levels. . . .

The Swedish welfare state is in a dilemma. Its ideology pushes for more and more government spending. . . . But its citizens reach a saturation point after which further tax increases are resisted. . . . the only ways Swedes can resist the higher taxes is by acting in ways detrimental to the economy. Rising public expenditures thereby undercut the economic base upon which the welfare economy depends.

WHY SPECIAL INTERESTS PREVAIL

If the cresting of the tide toward Fabian socialism and New Deal liberalism is to be followed by a move toward a freer society and a more limited government rather than toward a totalitarian society, the public must not only recognize the defects of the present situation but also how it has come about and what we can do about it. Why are the results of policies so often the opposite of their ostensible objectives? Why do special interests prevail over the general interest? What devices can we use to stop and reverse the process?

The Power in Washington

Whenever we visit Washington, D.C., we are impressed all over again with how much power is concentrated in that city. Walk the halls of Congress, and the 435 members of the House plus the 100 senators are hard to find among their 18,000 employees —about 65 for each senator and 27 for each member of the House. In addition, the more than 15,000 registered lobbyists—often accompanied by secretaries, typists, researchers, or representatives of the special interest they represent—walk the same halls seeking to exercise influence.

And this is but the tip of the iceberg. The federal government employs close to 3 million civilians (excluding the uniformed military forces). Over 350,000 are in Washington and the surrounding metropolitan area. Countless others are indirectly employed through government contracts with nominally private organizations, or are employed by labor or business organizations or other special interest groups that maintain their headquarters, or at least an office, in Washington because it is the seat of government.

Washington is a magnet for lawyers. Many of the country's largest and most affluent firms are located there. There are said to be more than 7,000 lawyers in Washington engaged in federal or regulatory practice alone. Over 160 out-of-town law firms have Washington offices.[3]

The power in Washington is not monolithic power in a few hands, as it is in totalitarian countries like the Soviet Union or Red China or, closer to home, Cuba. It is fragmented into many bits and pieces. Every special group around the country tries to get its hands on whatever bits and pieces it can. The result is that there is hardly an issue on which government is not on both sides.

For example, in one massive building in Washington some government employees are working full-time trying to devise and implement plans to spend our money to discourage us from smoking cigarettes. In another massive building, perhaps miles away from the first, other employees, equally dedicated, equally hardworking, are working full-time spending our money to subsidize farmers to grow tobacco.

In one building the Council on Wage and Price Stability is working overtime trying to persuade, pressure, hornswoggle businessmen to hold down prices and workers to restrain their wage demands. In another building some subordinate agencies in the Department of Agriculture are administering programs to keep up, or raise, the prices of sugar, cotton, and numerous other agricultural products. In still another building officials of the Department of Labor are making determinations of "prevailing wages" under the Davis-Bacon Act that are pushing up the wage rates of construction workers.

Congress set up a Department of Energy employing 20,000 persons to promote the conservation of energy. It also set up an Environmental Protection Agency employing over 12,000 persons to issue regulations and orders, most of which require the use of more energy. No doubt, within each agency there are subgroups working at cross-purposes.

The situation would be ludicrous if it were not so serious. While many of these effects cancel out, their costs do not. Each program takes money from our pockets that we could use to buy goods and services to meet our separate needs. Each of them uses able, skilled people who could be engaged in productive activities. Each one grinds out rules, regulations, red tape, forms to fill in that bedevil us all.

Concentrated versus Diffuse Interests

Both the fragmentation of power and the conflicting government policies are rooted in the political realities of a democratic system that operates by enacting detailed and specific legislation. Such a system tends to give undue political power to small groups that have highly concentrated interests, to give greater weight to obvious, direct, and immediate effects of government action than to possibly more important but concealed, indirect, and delayed effects, to set in motion a process that sacrifices the general interest to serve special interests, rather than the other way around. There is, as it were, an invisible hand in politics that operates in precisely the opposite direction to Adam Smith's invisible hand. Individuals who intend only to promote the *general interest* are led by the invisible political hand to promote a *special interest* that they had no intention to promote.

A few examples will clarify the nature of the problem. Consider the government program of favoring the merchant marine by subsidies for shipbuilding and operations and by restricting much coastal traffic to American-flag ships. The estimated cost to the taxpayer is about $600 million a year—or $15,000 per year for each of the 40,000 people actively engaged in the industry. Ship owners, operators, and their employees have a strong incentive to get and keep those measures. They spend money lavishly for lobbying and political contributions. On the other hand, $600 million divided by a population of over 200 million persons comes to $3 a person per year; $12 for a family of four. Which of us will vote against a candidate for Congress because he imposed that cost on us? How many of us will deem it worth spending money to defeat such measures, or even spending time to become informed about such matters?

As another example, the owners of stock in steel companies, the executives of these companies, the steelworkers all know very well that an increase in the importation of foreign steel into the United States will mean less money and fewer jobs for them. They clearly recognize that government action to keep out imports will benefit them. Workers in export industries who will lose their jobs because fewer imports from Japan mean fewer exports to Japan do not know that their jobs are threatened. When they lose their jobs, they do not know why. The purchasers of automobiles or of kitchen stoves or of other items made of steel may complain about the higher prices they have to pay. How many purchasers will trace the higher price back to the restriction on steel imports that forces manufacturers to use higher-priced domestic steel instead of lower-priced foreign steel? They are far more likely to blame "greedy" manufacturers or "grasping" trade unionists.

Agriculture is another example. Farmers descend on Washington in their tractors to demonstrate for higher price supports. Before the change in the role of government that made it natural to appeal to Washington, they would have blamed the bad weather and repaired to churches, not the White House, for assistance. Even for so indispensable and visible a product as food, no consumers parade in Washington to protest the price supports. And the farmers themselves, even though agriculture is the major export industry of the United States, do not recognize the extent to which their own problems arise from government's interference with foreign trade. It never occurs to them, for example, that they may be harmed by restrictions on steel imports.

Or to take a very different example, the U.S. Post Office. Every movement to remove the government monopoly of first-class mail is vigorously opposed by the trade unions of postal workers. They recognize very clearly that opening postal service to private enterprise may mean the loss of their jobs. It pays them to try to prevent that outcome. As the case of the Brennans in Rochester suggests, if the postal monopoly were abolished, a vigorous private industry would arise, containing thousands of firms and employing tens of thousands of workers. Few of the people who might find a rewarding opportunity in such an industry even know that the possibility exists. They are certainly not in Washington testifying to the relevant congressional committee.

The benefit an individual gets from any one program that he has a special interest in may be more than canceled by the costs to him of many programs that affect him lightly. Yet it pays him to favor the one program, and not oppose the others. He can readily recognize that he and the small group with the same special interest can afford to spend enough money and time to make a difference in respect of the one program. Not promoting that program will not prevent the others, which do him harm, from being adopted. To achieve that, he would have to be willing and able to devote as much effort to opposing each of them as he does to favoring his own. That is clearly a losing proposition.

Citizens are aware of taxes—but even that awareness is diffused by the hidden nature of most taxes. Corporate and excise taxes are paid for in the prices of the goods people buy, without separate accounting. Most income taxes are withheld at source. Inflation, the worst of the hidden taxes, defies easy understanding. Only sales taxes, property taxes, and income taxes in excess of withholding are directly and painfully visible—and they are the taxes on which resentment centers.

Bureaucracy

The smaller the unit of government and the more restricted the functions assigned government, the less likely it is that its actions will reflect special interests rather than the general interest. The New England town meeting is the image that comes to mind. The people governed know and can control the people governing; each person can express his views; the agenda is sufficiently small that everyone can be reasonably well informed about minor items as well as major ones.

As the scope and role of government expands—whether by covering a larger area and population or by performing a wider variety of functions—the connection between the people governed and the people governing becomes attenuated. It becomes impossible for any large fraction of the citizens to be reasonably well informed about all items on the vastly enlarged government agenda, and, beyond a point, even about all major items. The bureaucracy that is needed to administer government grows and increasingly interposes itself between the citizenry and the representatives they choose. It becomes both a vehicle whereby special interests can achieve their objectives and an important special interest in its own right—a major part of the new class referred to in Chapter 5.

Currently in the United States, anything like effective detailed control of government by the public is limited to villages, towns, smaller cities, and suburban areas—and even there only to those matters not mandated by the state or federal government. In large cities, states, Washington, we have government of the people not by the people but by a largely faceless group of bureaucrats.

No federal legislator could conceivably even read, let alone analyze and study, all the laws on which he must vote. He must depend on his numerous aides and assistants, or outside lobbyists, or fellow legislators, or some other source for most of his decisions on how to vote. The unelected congressional bureaucracy almost surely has far more influence today in shaping the detailed laws that are passed than do our elected representatives.

The situation is even more extreme in the administration of government programs. The vast federal bureaucracy spread through the many government departments and independent agencies is literally out of control of the elected representatives of the public. Elected Presidents and senators and representatives come and go but the civil service remains. Higher-level bureaucrats are past masters at the art of using red tape to delay and defeat proposals they do not favor; of issuing rules and regulations as "interpretations" of laws that in fact subtly, or sometimes crudely, alter their thrust; of dragging their feet in administering those parts of laws of which they disapprove, while pressing on with those they favor.

More recently, the federal courts, faced with increasingly complex and far-reaching legislation, have departed from their traditional role as impersonal interpreters of the law and have become active participants in both legislation and administration. In doing so, they have become part of the bureaucracy rather than an independent part of the government mediating between the other branches.

Bureaucrats have not usurped power. They have not deliberately engaged in any kind of conspiracy to subvert the democratic process. Power has been thrust on them. It is simply impossible to conduct complex government activities in any other way than by delegating responsibility. When that leads to conflicts between bureaucrats delegated different functions—as, recently, between bureaucrats instructed to preserve and improve the environment and bureaucrats instructed to foster the conservation and production of energy—the only solution that is available is to give power to another set of bureaucrats to resolve the conflict—to cut red tape, it is said, when the real problem is not red tape but a conflict between desirable objectives.

The high-level bureaucrats who have been assigned these functions cannot imagine that the reports they write or receive, the meetings they attend, the lengthy discussions they hold with other important people, the rules and regulations they issue—that all these are the problem rather than the solution. They inevitably become persuaded that they are indispensable, that they know more about what should be done than uninformed voters or self-interested businessmen.

The growth of the bureaucracy in size and power affects every detail of the relation between a citizen and his government. If you have a grievance or can see a way of gaining an advantage from a government measure, your first recourse these days is likely to be to try to influence a bureaucrat to rule in your favor. You may appeal to your elected representative, but if so, you are perhaps more likely to ask him to intervene on your behalf with a bureaucrat than to ask him to support a specific piece of legislation.

Increasingly, success in business depends on knowing one's way around Washington, having influence with legislators and bureaucrats. What has come to be called a "revolving door" has developed between government and business. Serving a term as a civil servant in Washington has become an apprenticeship for a successful business career. Government jobs are sought less as the first step in a lifetime government career than for the value of contacts and inside knowledge to a possible future employer. Conflict-of-interest legislation proliferates, but at best only eliminates the most obvious abuses.

When a special interest seeks benefits through highly visible legislation, it not only must clothe its appeal in the rhetoric of the general interest, it must persuade a significant segment of disinterested persons that its appeal has merit. Legislation recognized as naked self-interest will seldom be adopted—as illustrated by the recent defeat of further special privileges to the merchant marine despite endorsement by President Carter after receiving substantial campaign assistance from the unions involved. Protecting the steel industry from foreign competition is promoted as contributing to national security and full employment; subsidizing agriculture as assuring a reliable supply of food; the postal monopoly as cementing the nation together; and so on without end.

Nearly a century ago, A. V. Dicey explained why the rhetoric in terms of the general interest is so persuasive: "The beneficial effect of state intervention, especially in the form of legislation, is direct, immediate, and so to speak, visible, while its evil effects are gradual and indirect, and lie out of sight. . . . Hence the majority of mankind must almost of necessity look with undue favor upon governmental intervention." [4]

This "natural bias," as he termed it, in favor of government intervention is enormously strengthened when a special interest seeks benefits through administrative procedures rather than legislation. A trucking company that appeals to the ICC for a favorable ruling also uses the rhetoric of the general interest, but no one is likely to press it on that point. The company need persuade no one except the bureaucrats. Opposition seldom comes from disinterested persons concerned with the general interest. It comes from other interested parties, shippers or other truckers, who have their own axes to grind. The camouflage wears very thin indeed.

The growth of the bureaucracy, reinforced by the changing role of the courts, has made a mockery of the ideal expressed by John Adams in his original (1779) draft of the Massachusetts constitution: "a government of laws instead of men." Anyone who has been subjected to a thorough customs inspection on returning from a trip abroad, had his tax returns audited by the Internal Revenue Service, been subject to inspection by an official of OSHA or any of a large number of federal agencies, had occasion to appeal to the bureaucracy for a ruling or a permit, or had to defend a higher price or wage before the Council on Wage and Price Stability is aware of how far we have come from a rule of law. The government official is supposed to be our servant. When you sit across the desk from a representative of the Internal Revenue Service who is auditing your tax return, which one of you is the master and which the servant?

Or to use a different illustration. A recent *Wall Street Journal* story (June 25, 1979) is headlined: "SEC's Charges Settled by a Former Director" of a corporation. The former director, Maurice G. McGill, is reported as saying, "The question wasn't whether I had personally benefited from the transaction but rather what the responsibilities of an outside director are. It would be interesting to take it to trial but my decision to settle was purely economic. The cost of fighting the SEC to completion would be enormous." Win or lose, Mr. McGill would have had to pay his legal costs. Win or lose, the SEC official prosecuting the case had little at stake except status among fellow bureaucrats.

WHAT WE CAN DO

Needless to say, those of us who want to halt and reverse the recent trend should oppose additional specific measures to expand further the power and scope of government, urge repeal and reform of existing measures, and try to elect legislators and executives who share that view. But that is not an effective way to reverse the growth of government. It is doomed to failure. Each of us would defend our own special privileges and try to limit government at someone else's expense. We would be fighting a many-headed hydra that would grow new heads faster than we could cut old ones off.

Our founding fathers have shown us a more promising way to proceed: by package deals, as it were. We should adopt self-denying ordinances that limit the objectives we try to pursue through political channels. We should not consider each case on its merits, but lay down broad rules limiting what government may do.

The merit of this approach is well illustrated by the First Amendment to the Constitution. Many specific restrictions on freedom of speech would be approved by a substantial majority of both legislators and voters. A majority would very likely favor preventing Nazis, Seventh-Day Adventists, Jehovah's Witnesses, the Ku Klux Klan, vegetarians, or almost any other little group you might name from speaking on a street corner.

The wisdom of the First Amendment is that it treats these cases as a bundle. It adopts the general principle that "Congress shall make no law . . . abridging the freedom of speech"; no consideration of each case on its merits. A majority supported it then and, we are persuaded, a majority would support it today. Each of us feels more deeply about not having our freedom interfered with when we are in the minority than we do about interfering with the freedom of others when we are in a majority—and a majority of us will at one time or another be in some minority.

We need, in our opinion, the equivalent of the First Amendment to limit government power in the economic and social area —an economic Bill of Rights to complement and reinforce the original Bill of Rights.

The incorporation of such a Bill of Rights into our Constitution would not in and of itself reverse the trend toward bigger government or prevent it from being resumed—any more than the original Constitution has prevented both a growth and a centralization of government power far beyond anything the framers intended or envisioned. A written constitution is neither necessary nor sufficient to develop or preserve a free society. Although Great Britain has always had only an "unwritten" constitution, it developed a free society. Many Latin American countries that adopted written constitutions copied from the United States Constitution practically word for word have not succeeded in establishing a free society. In order for a written—or for that matter, unwritten—constitution to be effective it must be supported by the general climate of opinion, among both the public at large and its leaders. It must incorporate principles that they have come to believe in deeply, so that it is taken for granted that the executive, the legislature, and the courts will behave in conformity to these principles. As we have seen, when that climate of opinion changes, so will policy.

Nonetheless, we believe that the formulation and adoption of an economic Bill of Rights would be the most effective step that could be taken to reverse the trend toward ever bigger government for two reasons: first, because the process of formulating the amendments would have great value in shaping the climate of opinion; second, because the enactment of amendments is a more direct and effective way of converting that climate of opinion into actual policy than our present legislative process.

Given that the tide of opinion in favor of New Deal liberalism has crested, the national debate that would be generated in formulating such a Bill of Rights would help to assure that opinion turned definitely toward freedom rather than toward totalitarianism. It would disseminate a better understanding of the problem of big government and of possible cures.

The political process involved in the adoption of such amendments would be more democratic, in the sense of enabling the values of the public at large to determine the outcome, than our present legislative and administrative structure. On issue after issue the government of the people acts in ways that the bulk of the people oppose. Every public opinion poll shows that a large majority of the public opposes compulsory busing for integrating schools—yet busing not only continues but is continuously expanded. Very much the same thing is true of affirmative action programs in employment and higher education and of many other measures directed at implementing views favorable to equality of outcome. So far as we know, no pollster has asked the public, "Are you getting your money's worth for the more than 40 percent of your income being spent on your behalf by government?" But is there any doubt what the poll would show?

For the reasons outlined in the preceding section, the special interests prevail at the expense of the general interest. The new class, enshrined in the universities, the news media, and especially the federal bureaucracy, has become one of the most powerful of the special interests. The new class has repeatedly succeeded in imposing its views, despite widespread public objection, and often despite specific legislative enactments to the contrary.

The adoption of amendments has the great virtue of being decentralized. It requires separate action in three-quarters of the states. Even the proposal of new amendments can bypass Congress: Article V of the Constitution provides that the "Congress . . . on the application of the Legislatures of two-thirds of the several states, shall call a convention for proposing amendments." The recent movement to call a convention to propose an amendment requiring the federal budget to be balanced was backed by thirty states by mid-1979. The possibility that four more state legislatures would join the move, making the necessary two-thirds, has sown consternation in Washington—precisely because it is the one device that can effectively bypass the Washington bureaucracy.

TAX AND SPENDING LIMITATIONS

The movement to adopt constitutional amendments to limit government is already under way in one area—taxes and spending. By early 1979 five states· had already adopted amendments to their constitutions that limit the amount of taxes that the state may impose, or in some cases the amount that the state may spend. Similar amendments are partway through the adoption process in other states and were scheduled to be voted on in still other states at the 1979 election. Active movements to have similar amendments adopted are under way in more than half the remaining states. A national organization, the National Tax Limitation Committee (NTLC), with which we are connected, has served as a clearinghouse and coordinator of the activities in the several states. It had about 250,000 members nationwide in mid-1979, and the number was climbing rapidly.

On the national level two important developments are under way. One is the drive to get state legislatures to mandate Congress to call a national convention to propose an amendment to balance the budget—sparked primarily by the National Taxpayers Union, which had over 125,000 members nationwide in mid-1979. The other is an amendment to limit spending at the federal level, which was drafted under the sponsorship of the NTLC. The drafting committee, on which we both served, included lawyers, economists, political scientists, state legislators, businessmen, and representatives of various organizations. The amendment it drafted has been introduced into both houses of Congress, and the NTLC is undertaking a national campaign in support of it. A copy of the proposed amendment is contained in Appendix B.

The basic idea behind both the state and federal amendments is to correct the defect in our present structure under which democratically elected representatives vote larger expenditures than a majority of voters deem desirable.

As we have seen, that outcome results from a political bias in favor of special interests. Government budgets are determined by adding together expenditures that are authorized for a host of separate programs. The small number of people who have a special interest in each specific program spend money and work hard to get it passed; the large number of people, each of whom will be assessed a few dollars to pay for the program, will not find it worthwhile to spend money or work to oppose it, even if they manage to find out about it.

The majority does rule. But it is a rather special kind of majority. It consists of a coalition of special interest minorities. The way to get elected to Congress is to collect groups of, say, 2 or 3 percent of your constituents, each of which is strongly interested in one special issue that hardly concerns the rest of your constituents. Each group will be willing to vote for you if you promise to back its issue regardless of what you do about other issues. Put together enough such groups and you will have a 51 percent majority. That is the kind of logrolling majority that rules the country.

The proposed amendments would alter the conditions under which legislators—state or federal, as the case may be—operate by limiting the total amount they are authorized to appropriate. The amendments would give the government a limited budget, specified in advance, the way each of us has a limited budget. Much special interest legislation is undesirable, but it is never clearly and unmistakably bad. On the contrary, every measure will be represented as serving a good cause. The problem is that there are an infinite number of good causes. Currently, a legislator is in a weak position to oppose a "good" cause. If he objects that it will raise taxes, he will be labeled a reactionary who is willing to sacrifice human need for base mercenary reasons—after all, this good cause will only require raising taxes by a few cents or dollars per person. The legislator is in a far better position if he can say, "Yes, yours is a good cause, but we have a fixed budget. More money for your cause means less for others. Which of these others should be cut?" The effect would be to require the special interests to compete with one another for a bigger share of a fixed pie, instead of their being able to collude with one another to make the pie bigger at the expense of the taxpayer.

Because states do not have the power to print money, state budgets can be limited by limiting total taxes that may be imposed, and that is the method that has been used in most of the state amendments that have been adopted or proposed. The federal government can print money, so limiting taxes is not an effective method. That is why our amendment is stated in terms of limiting total spending by the federal government, however financed.

The limits—on either taxes or spending—are mostly specified in terms of the total income of the state or nation in such a way that if spending equaled the limit, government spending would remain constant as a fraction of income. That would halt the trend toward ever bigger government, not reverse it. However, the limits would encourage a reversal because, in most cases, if spending did not equal the limit in any year, that would lower the limits applicable to future years. In addition, the proposed federal amendment requires a reduction in the percentage if inflation exceeds 3 percent a year.

OTHER CONSTITUTIONAL PROVISIONS

A gradual reduction in the fraction of our income that government spends would be a major contribution to a freer and stronger society. But it would be only one step toward that objective.

Many of the most damaging kinds of government controls over our lives do not involve much government spending: for example, tariffs, price and wage controls, licensure of occupations, regulation of industry, consumer legislation.

With respect to these, too, the most promising approach is through general rules that limit government power. As yet, the designing of appropriate rules of this kind has received little attention. Before any rules can be taken seriously, they need the kind of thorough examination by people with different interests and knowledge that the tax and spending limitation amendments have received.

As a first step in this process, we sketch a few examples of the kinds of amendments that appear to us desirable. We stress that these are highly tentative, intended primarily to stimulate further thought and further work in this largely unexplored area.

International Trade

The Constitution now specifies, "No State shall, without the consent of the Congress, lay any imposts or duties on imports or exports, except what may be absolutely necessary for executing its inspection laws." An amendment could specify:

> *Congress shall not lay any imposts or duties on imports or exports, except what may be absolutely necessary for executing its inspection laws.*

It is visionary to suppose that such an amendment could be enacted now. However, achieving free trade through repealing individual tariffs is, if anything, even more visionary. And the attack on all tariffs consolidates the interests we all have as consumers to counter the special interest we each have as producers.

Wage and Price Controls

As one of us wrote some years ago, "If the U.S. ever succumbs to collectivism, to government control over every facet of our lives, it will not be because the socialists win any arguments. It will be through the indirect route of wage and price controls." [5] Prices, as we noted in Chapter 1, transmit information—which Walter Wriston has quite properly translated by describing prices as a form of speech. And prices determined in a free market are a form of free speech. We need here the exact counterpart of the First Amendment:

> *Congress shall make no laws abridging the freedom of sellers of goods or labor to price their products or services.*

Occupational Licensure

Few things have a greater effect on our lives than the occupations we may follow. Widening freedom to choose in this area requires limiting the power of states. The counterpart here in our Constitution is either the provisions in its text which prohibit certain actions by states or the Fourteenth Amendment. One suggestion:

> *No State shall make or impose any law which shall abridge the right of any citizen of the United States to follow any occupation or profession of his choice.*

A Portmanteau Free Trade Amendment

The three preceding amendments could all be replaced by a single amendment patterned after the Second Amendment to our Constitution (which guarantees the right to keep and bear arms):

> *The right of the people to buy and sell legitimate goods and services at mutually acceptable terms shall not be infringed by Congress or any of the States.*

Taxation

By general consent, the personal income tax is sadly in need of reform. It professes to adjust the tax to "ability to pay," to tax the rich more heavily and the poor less heavily and to allow for each individual's special circumstances. It does no such thing. Tax rates are highly graduated on paper, rising from 14 to 70 percent. But the law is riddled with so many loopholes, so many special privileges, that the high rates are almost pure window dressing. A low flat rate—less than 20 percent—on all income above personal exemptions with no deductions except for strict occupational expenses would yield more revenue than the present unwieldy structure. Taxpayers would be better off—because they would be spared the costs of sheltering income from taxes; the economy would be better off—because tax considerations would play a smaller role in the allocation of resources. The only losers would be lawyers, accountants, civil servants, and legislators—who would have to turn to more productive activities than filling in tax forms, devising tax loopholes, and trying to close them.

The corporate income tax, too, is highly defective. It is a hidden tax that the public pays in the prices it pays for goods and services without realizing it. It constitutes double taxation of corporate income—once to the corporation, once to the stockholder when the income is distributed. It penalizes capital investment and thereby hinders growth in productivity. It should be abolished.

Although there is agreement between left and right that lower rates, fewer loopholes, and a reduction in the double taxation of corporate income would be desirable, such a reform cannot be enacted through the legislative process. The left fear that if they accepted lower rates and less graduation in return for eliminating loopholes, new loopholes would soon emerge—and they are right. The right fear that if they accepted the elimination of the loopholes in return for lower rates and less graduation, steeper graduation would soon emerge—and they are right.

This is a specially clear case where a constitutional amendment is the only hope of striking a bargain that all sides can expect to be honored. The amendment needed here is the repeal of the present Sixteenth Amendment authorizing income taxes and its replacement by one along the following lines:

The Congress shall have power to lay and collect taxes on incomes of persons, from whatever sources derived, without apportionment among the several States, and without regard to any census or enumeration, provided that the same tax rate is applied to all income in excess of occupational and business expenses and a personal allowance of a fixed amount. The word "person" shall exclude corporations and other artificial persons.

Sound Money

When the Constitution was enacted, the power given to Congress "to coin money, regulate the value thereof, and of foreign coin" referred to a commodity money: specifying that the dollar shall mean a definite weight in grams of silver or gold. The paper money inflation during the Revolution, as well as earlier in various colonies, led the framers to deny states the power to "coin money; emit bills of credit [i.e., paper money]; make anything but gold and silver coin a tender in payment of debts." The Constitution is silent on Congress's power to authorize the government to issue paper money. It was widely believed that the Tenth Amendment, providing that the "powers not delegated to the United States by the Constitution . . . are reserved to the States respectively, or to the people," made the issuance of paper money unconstitutional.

During the Civil War, Congress authorized greenbacks and made them a legal tender for all debts public and private. After the Civil War, in the first of the famous greenback cases, the Supreme Court declared the issuance of greenbacks unconstitutional. One "fascinating aspect of this decision is that it was delivered by Chief Justice Salmon P. Chase, who had been Secretary of the Treasury when the first greenbacks were issued. Not only did he not disqualify himself, but in his capacity as Chief Justice convicted himself of having been responsible for an unconstitutional action in his capacity as Secretary of the Treasury." [6]

Subsequently an enlarged and reconstituted Court reversed the first decision by a majority of five to four, affirming that making greenbacks a legal tender was constitutional, with Chief Justice Chase as one of the dissenting justices.

It is neither feasible nor desirable to restore a gold- or silver-coin standard, but we do need a commitment to sound money. The best arrangement currently would be to require the monetary authorities to keep the percentage rate of growth of the monetary base within a fixed range. This is a particularly difficult amendment to draft because it is so closely linked to the particular institutional structure. One version would be:

> *Congress shall have the power to authorize non-interest-bearing obligations of the government in the form of currency or book entries, provided that the total dollar amount outstanding increases by no more than 5 percent per year and no less than 3 percent.*

It might be desirable to include a provision that two-thirds of each House of Congress, or some similar qualified majority, can waive this requirement in case of a declaration of war, the suspension to terminate annually unless renewed.

Inflation Protection

If the preceding amendment were adopted and strictly adhered to, that would end inflation and assure a relatively stable price level. In that case, no further measures would be needed to prevent the government from engaging in inflationary taxation without representation. However, that is a big *if*. An amendment that would remove the incentive for government to inflate would have broad support. It might be adopted far more readily than a more technical and controversial sound-money amendment. In effect, what is required is the extension of the Fifth Amendment provision that "[n]o person shall . . . be deprived of life, liberty, or property, without due process of law; nor shall private property be taken for public use without just compensation."

A person whose dollar income just keeps pace with inflation yet who is pushed into a higher tax bracket is deprived of property without due process. The repudiation of part of the real value of government bonds through inflation is the taking of private property for public use without just compensation.

The relevant amendment would specify:

All contracts between the U.S. government and other parties stated in dollars, and all dollar sums contained in federal laws, shall be adjusted annually to allow for the change in the general level of prices during the prior year.

Like the monetary amendment, this, too, is difficult to draft precisely because of its technical character. Congress would have to specify precise procedures, including what index number should be used to approximate "the general level of prices." But it states the fundamental principle.

This is hardly an exhaustive list—we still have three to go to match the ten amendments in the original Bill of Rights. And the suggested wording needs the scrutiny of experts in each area as well as constitutional legal experts. But we trust that these proposals at least indicate the promise of a constitutional approach.

CONCLUSION

The two ideas of human freedom and economic freedom working together came to their greatest fruition in the United States. Those ideas are still very much with us. We are all of us imbued with them. They are part of the very fabric of our being. But we have been straying from them. We have been forgetting the basic truth that the greatest threat to human freedom is the concentration of power, whether in the hands of government or anyone else. We have persuaded ourselves that it is safe to grant power, provided it is for good purposes.

Fortunately, we are waking up. We are again recognizing the dangers of an overgoverned society, coming to understand that good objectives can be perverted by bad means, that reliance on the freedom of people to control their own lives in accordance with their own values is the surest way to achieve the full potential of a great society.

Fortunately, also, we are as a people still free to choose which way we should go—whether to continue along the road we have been following to ever bigger government, or to call a halt and change direction.

APPENDICES

APPENDIX A

SOCIALIST PLATFORM OF 1928

Herewith the economic planks of the Socialist party platform of 1928, along with an indication in parentheses of how these planks have fared. The list that follows includes every economic plank, but not the full language of each.

1. "Nationalization of our natural resources, beginning with the coal mines and water sites, particularly at Boulder Dam and Muscle Shoals." (Boulder Dam, renamed Hoover Dam, and Muscle Shoals are now both federal government projects.)

2. "A publicly owned giant power system under which the federal government shall cooperate with the states and municipalities in the distribution of electrical energy to the people at cost." (Tennessee Valley Authority.)

3. "National ownership and democratic management of railroads and other means of transportation and communication." (Railroad passenger service is completely nationalized through Amtrak. Some freight service is nationalized through Conrail. The FCC controls communications by telephone, telegraph, radio, and television.)

4. "An adequate national program for flood control, flood relief, reforestation, irrigation, and reclamation." (Government expenditures for these purposes are currently in the many billions of dollars.)

5. "Immediate governmental relief of the unemployed by the extension of all public works and a program of long range planning of public works . . ." (In the 1930s, WPA and PWA were a direct counterpart; now, a wide variety of other programs are.) "All persons thus employed to be engaged at hours and wages fixed by bona-fide labor unions." (The Davis-Bacon and Walsh-Healey Acts require contractors with government contracts to pay "prevailing wages," generally interpreted as highest union wages.)

6. "Loans to states and municipalities without interest for the purpose of carrying on public works and the taking of such other measures as will lessen widespread misery." (Federal grants in aid to states and local municipalities currently total tens of billions of dollars a year.)

7. "A system of unemployment insurance." (Part of Social Security system.)

8. "The nation-wide extension of public employment agencies in cooperation with city federations of labor." (U.S. Employment Service

and affiliated state employment services administer a network of about 2,500 local employment offices.)

9. "A system of health and accident insurance and of old age pensions as well as unemployment insurance." (Part of Social Security system.)

10. "Shortening the workday" and "Securing to every worker a rest period of no less than two days in each week." (Legislated by wages and hours laws that require overtime for more than forty hours of work per week.)

11. "Enacting of an adequate federal anti–child labor amendment." (Not achieved as amendment, but essence incorporated in various legislative acts.)

12. "Abolition of the brutal exploitation of convicts under the contract system and substituticn of a cooperative organization of industries in penitentiaries and workshops for the benefit of convicts and their dependents." (Partly achieved, partly not.)

13. "Increase of taxation on high income levels, of corporation taxes and inheritance taxes, the proceeds to be used for old age pensions and other forms of social insurance." (In 1928, highest personal income tax rate, 25 percent; in 1978, 70 percent; in 1928, corporate tax rate, 12 percent; in 1978, 48 percent; in 1928, top federal estate tax rate, 20 percent; in 1978, 70 percent.)

14. "Appropriation by taxation of the annual rental value of all land held for speculation." (Not achieved in this form, but property taxes have risen drastically.)

APPENDIX B

A PROPOSED CONSTITUTIONAL AMENDMENT
TO LIMIT FEDERAL SPENDING
Prepared by the Federal Amendment Drafting Committee
W. C. Stubblebine, Chairman
Convened by The National Tax Limitation Committee
Wm. F. Rickenbacker, Chairman; Lewis K. Uhler, President

Section 1. To protect the people against excessive governmental burdens and to promote sound fiscal and monetary policies, total outlays of the Government of the United States shall be limited.

(a) Total outlays in any fiscal year shall not increase by a percentage greater than the percentage increase in nominal gross national product in the last calendar year ending prior to the beginning of said fiscal year. Total outlays shall include budget and off-budget outlays, and exclude redemptions of the public debt and emergency outlays.

(b) If inflation for the last calendar year ending prior to the beginning of any fiscal year is more than three per cent, the permissible percentage increase in total outlays for that fiscal year shall be reduced by one-fourth of the excess of inflation over three per cent. Inflation shall be measured by the difference between the percentage increase in nominal gross national product and the percentage increase in real gross national product.

Section 2. When, for any fiscal year, total revenues received by the Government of the United States exceed total outlays, the surplus shall be used to reduce the public debt of the United States until such debt is eliminated.

Section 3. Following declaration of an emergency by the President, Congress may authorize, by a two-thirds vote of both Houses, a specified amount of emergency outlays in excess of the limit for the current fiscal year.

Section 4. The limit on total outlays may be changed by a specified amount by a three-fourths vote of both Houses of Congress when approved by the Legislatures of a majority of the several States. The change shall become effective for the fiscal year following approval.

Section 5. For each of the first six fiscal years after ratification of this article, total grants to States and local governments shall not be a smaller fraction of total outlays than in the three fiscal years prior to the ratification of this article. Thereafter, if grants are less than that fraction of total outlays, the limit on total outlays shall be decreased by an equivalent amount.

Section 6. The Government of the United States shall not require, directly or indirectly, that States or local governments engage in additional or expanded activities without compensation equal to the necessary additional costs.

Section 7. This article may be enforced by one or more members of the Congress in an action brought in the United States District Court for the District of Columbia, and by no other persons. The action shall name as defendant the Treasurer of the United States, who shall have authority over outlays by any unit or agency of the Government of the United States when required by a court order enforcing the provisions of this article. The order of the court shall not specify the particular outlays to be made or reduced. Changes in outlays necessary to comply with the order of the court shall be made no later than the end of the third full fiscal year following the court order.

NOTES

INTRODUCTION

1. Adam Smith, *The Wealth of Nations* (1776). (All page references are to the edition edited by Edwin Cannan, 5th ed. (London: Methuen & Co., Ltd., 1930).
2. *On Liberty*, People's ed. (London: Longmans, Green & Co., 1865), p. 6.
3. *Wealth of Nations*, vol. I, p. 325 (Book II, Chap. III).

CHAPTER 1

1. See Hedrick Smith, *The Russians* (New York: Quadrangle Books/New York Times Book Co., 1976), and Robert G. Kaiser, *Russia: The People and the Power* (New York: Atheneum, 1976).
2. *Freeman*, December 1958.
3. *Wealth of Nations*, vol. II, pp. 184–85.

CHAPTER 2

1. *Wealth of Nations*, vol. I, pp. 422 and 458.
2. See George J. Stigler, *Five Lectures on Economic Problems* (New York: Macmillan, 1950), pp. 26–34.
3. "A New Holiday," *Newsweek*, August 5, 1974, p. 56.

CHAPTER 3

1. Lester V. Chandler, *Benjamin Strong, Central Banker* (Washington, D.C.: Brookings Institution, 1958), p. 465.

2. Milton Friedman and Anna J. Schwartz, *A Monetary History of the United States, 1867–1960* (Princeton: Princeton University Press, 1963), p. 310.
3. *The Memoirs of Herbert Hoover,* vol. III: *The Great Depression, 1929–1941* (New York: Macmillan, 1952), p. 212.
4. *Annual Report,* 1933, pp. 1 and 20–21.
5. For a fuller discussion see Friedman and Schwartz, *Monetary History,* pp. 362–419.

CHAPTER 4

1. It is worth quoting the whole sentence in which these words appear, because it is such an accurate description of the direction in which we are moving as well as a wholly unintentional indictment of the effect: "No man any more has any care for the morrow, either for himself or his children, for the nation guarantees the nurture, education, and comfortable maintenance of every citizen from the cradle to the grave." Edward Bellamy, *Looking Backward* (New York: Modern Library, 1917; original date of publication, 1887), p. 70.
2. *An Over-Governed Society* (New York: The Free Press, 1976), p. 235.
3. A. V. Dicey, *Lectures on the Relation between Law and Public Opinion in England during the Nineteenth Century,* 2d ed. (London: Macmillan, 1914), p. xxxv.
4. Ibid., pp. xxxvi–xxxvii.
5. Ibid., pp. xxxvii–xxxix.
6. Cecil Driver, *Tory Radical* (New York: Oxford University Press, 1946).
7. Quoted in Ken Auletta, *The Streets Were Paved with Gold* (New York: Random House, 1979), p. 255.
8. Ibid., p. 253.
9. These figures refer only to OASDHI and state unemployment insurance; they exclude railroad and public employee retirement, veterans' benefits, and workmen's compensation, treating these as part of compensation under voluntary employment contracts.
10. Social Security Administration, *Your Social Security,* Department of Health, Education and Welfare Publication No. (SSA) 77-10035 (June 1977), p. 24. The earliest version of the booklet we have seen is for 1969, but we conjecture that the booklet was first issued many years earlier. The words were changed in the February 1978 version, by which time the myth that "trust funds" played an important part had become transparent.
    The revised version reads: "The basic idea of social security is a

simple one: During working years, employees, their employers, and self-employed people pay social security contributions. This money is used only to pay benefits to the more than 33 million people getting benefits and to pay administrative costs of the program. Then, when today's worker's earnings stop or are reduced because of retirement, death, or disability, benefits will be paid to them from contributions by people in covered employment and self-employment at that time. These benefits are intended to replace part of the earnings the family has lost."

This is certainly a far more defensible statement, though it still labels "taxes" as "contributions." When we first discovered the change, we thought it might be a result of a *Newsweek* column one of us wrote in 1971 making the criticisms that follow in the text and repeated in a debate the same year with Wilbur J. Cohen, former Secretary of HEW. However, the delay of six years before the change was made exploded that conjecture.

11. George Orwell, *Nineteen Eighty-four* (New York: Harcourt Brace, 1949).

12. Social Security Administration, *Your Social Security,* Department of Health, Education and Welfare Publication No. (SSA) 79-10035 (January 1979), p. 5. This sentence was changed in 1973, the word "earning" replacing the words "now building."

13. J. A. Pechman, H. J. Aaron, and M. K. Taussig, *Social Security: Perspectives for Reform* (Washington, D.C.: Brookings Institution, 1968), p. 69.

14. John A. Brittain, *The Payroll Tax for Social Security* (Washington, D.C.: Brookings Institution, 1972).

15. George J. Stigler, "Director's Law of Public Income Redistribution," *Journal of Law and Economics,* vol. 13 (April 1970), p. 1.

16. See Martin Anderson, *Welfare* (Stanford, Calif.: Hoover Institution, Stanford University, 1978), Chap. 1, for an excellent discussion of the poverty estimates.

17. Ibid., p. 39.

18. Ibid., p. 91; based on his earlier book, *The Federal Bulldozer: A Critical Analysis of Urban Renewal, 1949–1962* (Cambridge, Mass.: The MIT Press, 1964).

19. "The FTC Discovers HUD," *Wall Street Journal,* March 21, 1979, p. 22.

20. From an unpublished paper, "How to Be a Clinician in a Socialist Country," given in 1976 at the University of Chicago.

21. Max Gammon, *Health and Security: Report on Public Provision for Medical Care in Great Britain* (London: St. Michael's Organization, December 1976), pp. 19, 18.

22. The elegant formulation as a two-by-two table arose out of discussions with Eben Wilson, an associate producer of our television program.

23. However, a recent innovation is that families with one or more de-

pendent children may qualify for a payment called an earned income credit, which is similar to a negative income tax.

24. There is a provision for averaging income over a number of years. But the conditions are fairly stringent, so a person with a fluctuating income pays more tax than a person with a stable income averaging the same amount. In addition, most people with fluctuating incomes do not benefit from it at all.

25. We proposed it in *Capitalism and Freedom* (Chicago: University of Chicago Press, 1962), Chap. 12; for Milton Friedman's testimony, see U.S. Congress, House, Committee on Ways and Means, *Social Security and Welfare Proposals, Hearings,* 91st Congress, 1st session, November 7, 1969, part 6, pp. 1944–1958.

26. For the role of the welfare bureaucracy in defeating President Nixon's plan, see Daniel P. Moynihan, *The Politics of a Guaranteed Income: The Nixon Administration and the Family Assistance Plan* (New York: Random House, 1973).

27. Anderson, *Welfare,* p. 135.

28. Ibid., p. 135.

29. Ibid., p. 142.

CHAPTER 5

1. See J. R. Pole, *The Pursuit of Equality in American History* (Berkeley and Los Angeles: University of California Press, 1978), pp. 51–58.

2. Alexis de Tocqueville, *Democracy in America,* 2 vols., 2d ed., trans. Henry Reeve, ed. Francis Bowen (Boston: John Allyn, Publisher, 1863), vol. I, pp. 66–67. (First French edition published in 1835.)

3. Ibid., pp. 67–68.

4. See Smith, *The Russians,* and Kaiser, *Russia: The People and the Power.* Nick Eberstadt, "Has China Failed?" *The New York Review of Books,* April 5, 1979, p. 37, notes, "In China, . . . income distribution seems *very roughly* to have been the same since 1953."

5. Helen Lefkowitz Horowitz, *Culture and the City* (Lexington: University Press of Kentucky, 1976), pp. ix–x.

6. Ibid., pp. 212 and 31.

7. "The Forgotten Man," in Albert G. Keller and Maurice R. Davis, eds., *Essays of William G. Sumner* (New Haven: Yale University Press, 1934), vol. I, pp. 466–96.

8. Robert Nozick, "Who Would Choose Socialism?" *Reason,* May 1978, pp. 22–23.

9. *Wealth of Nations,* vol. I, p. 325 (Book II, Chap. III).

10. See Smith, *The Russians,* and Kaiser, *Russia: The People and the Power.*

11. Nick Eberstadt, "China: How Much Success," *New York Review of Books,* May 3, 1979, pp. 40–41.
12. John Stuart Mill, *The Principles of Political Economy* (1848), 9th ed. (London: Longmans, Green & Co., 1886), vol. II, p. 332 (Book IV, Chap. VI).

CHAPTER 6

1. Leonard Billet, *The Free Market Approach to Educational Reform,* Rand Paper P-6141 (Santa Monica, Calif.: The Rand Corporation, 1978), pp. 27–28.
2. From *The Good Society,* as quoted by Wallis in *An Over-Governed Society,* p. viii.
3. Quoted by E. G. West, "The Political Economy of American Public School Legislation," *Journal of Law and Economics,* vol. 10 (October 1967), pp. 101–28, quotation from p. 106.
4. Ibid., p. 108.
5. Note the misleading terminology. "Public" is equated with "governmental," though in other contexts, as in "public utilities," "public libraries," and so on, that is not done. In schooling, is there any relevant sense in which Harvard College is less "public" than the University of Massachusetts?
6. Ibid., p. 110.
7. R. Freeman Butts, *Encyclopaedia Britannica,* vol. 7 (1970), p. 992.
8. W. O. L. Smith, *Encyclopaedia Britannica,* vol. 7 (1970), p. 988.
9. Ibid., pp. 988–89.
10. E. G. West, *Education and the State* (London: The Institute of Economic Affairs, 1965).
11. Gammon, *Health and Security,* p. 27.
12. We are indebted to Herbert Lobsenz and Cynthia Savo of Market Data Retrieval for making these data available to us from their Education Data Bank.
13. Indeed, many of these public schools can be regarded as, in effect, tax loopholes. If they were private, the tuition charges would not be deductible for purposes of the federal income tax. As public schools financed by local taxes, the taxes are deductible.
14. One of us first proposed this voucher plan in Milton Friedman, "The Role of Government in Education," in Robert A. Solo, ed., *Economics and the Public Interest* (New Brunswick, N.J.: Rutgers University Press, 1955). A revised version of this article is Chapter 6 of *Capitalism and Freedom.*
15. Ibid., p. 86.
16. See Christopher Jencks and associates, *Education Vouchers: A Report*

on Financing Elementary Education by Grants to Parents (Cambridge, Mass.: Center for the Study of Public Policy, December 1970); John E. Coons and Stephen D. Sugarman, *Education by Choice: The Case for Family Control* (Berkeley: University of California Press, 1978).

17. Coons and Sugarman, *Education by Choice,* p. 191.
18. Ibid., p. 130.
19. *Wealth of Nations,* vol. II, p. 253 (Book V, Chap. I).
20. For example, the Citizens for Educational Freedom, the National Association for Personal Rights in Education.
21. Education Voucher Institute, incorporated in May 1979 in Michigan.
22. Kenneth B. Clark, "Alternative Public School Systems," in the special issue on *Equal Educational Opportunity* of the *Harvard Educational Review,* vol. 38, no. 1 (Winter 1968), pp. 100–113; passage cited from pp. 110–11.
23. Daniel Weiler, *A Public School Voucher Demonstration: The First Year at Alum Rock,* Rand Report No. 1495 (Santa Monica, Calif.: The Rand Corporation, 1974).
24. Henry M. Levin, "Aspects of a Voucher Plan for Higher Education," Occasional Paper 72-7, School of Education, Stanford University, July 1972, p. 16.
25. Carnegie Commission on Higher Education, *Higher Education: Who Pays? Who Benefits? Who Should Pay?* (McGraw-Hill, June 1973), pp. 2–3.
26. Ibid., p. 4.
27. Ibid., p. 4.
28. Ibid., p. 15.
29. Carnegie Foundation for the Advancement of Teaching, *More than Survival: Prospects for Higher Education in a Period of Uncertainty* (San Francisco: Jossey Bass Publishers, 1975), p. 7.
30. Carnegie Commission, *Higher Education,* p. 176. We have not calculated the percentages in the text from the Carnegie table but from the source it cited, Table 14, U.S. Census Reports Series P-20 for 1971, no. 241, p. 40. In doing so, we found that the Carnegie report percentages are slightly in error.
    The figures we give are somewhat misleading because married students living with their spouses are classified by their own and their spouses' family income rather than by the income of their parents. If married students are omitted, the effect described is even greater: 22 percent of students from families with incomes of less than $5,000 attended private schools, 17 percent from families with incomes between $5,000 and $10,000, and 25 percent from families with incomes of $10,000 and over.
31. According to figures from the U.S. Bureau of the Census, of those persons between eighteen and twenty-four who were enrolled as undergraduates in public colleges in 1971, fewer than 14 percent came from

families with incomes below $5,000 a year, although more than 22 percent of all eighteen- to twenty-four-year-olds came from these low-income families. And 57 percent of those enrolled came from families with incomes above $10,000 a year, although fewer than 40 percent of eighteen- to twenty-four-year-olds came from these higher-income families.

Again, these figures are biased by the inclusion of married students with spouse present. Only 9 percent of other students enrolled in public colleges came from families with incomes below $5,000, although 18 percent of all such eighteen- to twenty-four-year-olds came from these low-income families. Nearly 65 percent of students of other marital status enrolled came from families with incomes of $10,000 or more, although only a bit over 50 percent of all such eighteen- to twenty-four-year-olds did.

Incidentally, in connection with this and the preceding note, it is noteworthy that the Carnegie Commission, in the summary report in which it refers to these figures, does not even mention that it combines indiscriminately the married and unmarried students, even though doing so clearly biases their results in the direction of understating the transfer of income from lower to higher incomes that is involved in governmental financing of higher education.

32. Douglas M. Windham made two estimates for 1967–68 for each of four income classes of the difference between the dollar value of the benefits received from public higher education and the cost incurred. The estimates showing the smaller transfer are as follows.

Income Class ($ per year)	Total Benefits	Total Costs	Net Cost (−) or Gain (+)
$ 0– 3,000	$10,419,600	$14,259,360	−$ 3,839,760
3,000– 5,000	20,296,320	28,979,110	− 8,682,790
5,000–10,000	70,395,980	82,518,780	− 12,122,800
10,000 and over	64,278,490	39,603,440	+ 24,675,050

Douglas M. Windham, *Education, Equality and Income Redistribution* (Lexington, Mass.: Heath Lexington Books, 1970), p. 43.

33. W. Lee Hansen and Burton A. Weisbrod, *Benefits, Costs, and Finance of Public Higher Education* (Chicago: Markom Publishing Co., 1969), p. 76, except that line 5 below was calculated by us. Note that the taxes in line 3, unlike the costs allowed for in Florida, include all taxes, not simply the taxes going to pay for higher education.

	All Families	Families without Children in California Public Higher Education	Families with Children in California Public Higher Education			
			Total	Junior College	State College	Univ. of Calif.
1. Average family income	$8,000	$7,900	$9,560	$8,800	$10,000	$12,000
2. Average higher education subsidy per year	—	0	880	720	1,400	1,700
3. Average total state and local taxes paid	620	650	740	680	770	910
4. Net transfer (line 2 — line 3)	—	— 650	+ 140	+ 40	+ 630	+ 790
5. Net transfer as percent of average income		—8.2%	+ 1.5	+ 0.5	+ 6.3	+ 6.6

34. Carnegie Commission, *Higher Education*, p. 7.
35. Originally published in Milton Friedman, "The Role of Government in Education," and reprinted in slightly revised form in *Capitalism and Freedom;* quotation from p. 105 of the latter.
36. *Educational Opportunity Bank*, a Report of the Panel on Educational Innovation to the U.S. Commissioner of Education and the Director of the National Science Foundation (Washington, D.C.: U.S. Government Printing Office, August 1967). Supporting material was presented in K. Shell, F. M. Fisher, D. K. Foley, A. F. Friedlaender (in association with J. Behr, S. Fischer, K. Mosenson), "The Educational Opportunity Bank: An Economic Analysis of a Contingent Repayment Loan Program for Higher Education," *National Tax Journal*, March 1968, pp. 2–45, as well as in unpublished documents of the Zacharias Panel.
37. For the statement of the association, see National Association of State Universities and Land Grant Colleges, *Proceedings, November 12–15, 1967*, pp. 67–68. For the Smith quotation, *Wealth of Nations*, vol. I, p. 460 (Book IV, Chap III), where the reference is to traders seeking government protection from foreign goods.
38. Carnegie Commission, *Higher Education*, p. 121.
39. Quoted from *Capitalism and Freedom*, pp. 99–100.

CHAPTER 7

1. Marcia B. Wallace and Ronald J. Penoyer, "Directory of Federal Regulatory Agencies," Working Paper No. 36, Center for the Study of American Business, Washington University, St. Louis, September 1978, p. ii.

2. *Evaluation of the 1960–1963 Corvair Handling and Stability* (Washington, D.C.: U.S. Department of Transportation, National Highway Traffic Safety Administration, July 1972), p. 2.

3. See Mary Bennett Peterson, *The Regulated Consumer* (Los Angeles: Nash Publishing, 1971), p. 164.

4. Matthew Josephson, *The Politicos* (New York: Harcourt Brace, 1938), p. 526.

5. Thomas Gale Moore, "The Beneficiaries of Trucking Regulation," *Journal of Law and Economics*, vol. 21 (October 1978), p. 340.

6. Ibid., pp. 340, 342.

7. Gabriel Kolko, *The Triumph of Conservatism* (The Free Press of Glencoe, 1963), quotation from p. 99.

8. Richard Harris, *The Real Voice* (New York: Macmillan, 1964), p. 183.

9. William M. Wardell and Louis Lasagna, *Regulation and Drug Development* (Washington, D.C.: American Enterprise Institute for Public Policy Research, 1975), p. 8.

10. Sam Peltzman, *Regulation of Pharmaceutical Innovation* (Washington, D.C.: American Enterprise Institute for Public Policy Research, 1974), p. 9.

11. Estimates for 1950s and early 1960s from Wardell and Lasagna, *Regulation and Drug Development*, p. 46; for 1978, from Louis Lasagna, "The Uncertain Future of Drug Development," *Drug Intelligence and Clinical Pharmacy*, vol. 13 (April 1979), p. 193.

12. Peltzman, *Regulation of Pharmaceutical Innovation*, p. 45.

13. U.S. Consumer Products Safety Commission, *Annual Report, Fiscal Year 1977* (Washington, D.C., January 1978), p. 4.

14. Wallace and Penoyer, "Directory of Federal Regulatory Agencies," p. 14.

15. Murray L. Weidenbaum, *The Costs of Government Regulation*, Publication No. 12 (St. Louis: Center for the Study of American Business, Washington University, February 1977), p. 9.

16. Ibid.

17. Wallace and Penoyer, "Directory of Federal Regulatory Agencies," p. 19.

18. A. Myrick Freeman III and Ralph H. Haveman, "Clean Rhetoric and Dirty Water," *The Public Interest,* No. 28 (Summer 1972), p. 65.
19. Herbert Asbury, *The Great Illusion, An Informal History of Prohibition* (Garden City, N.Y.: Doubleday, 1950), pp. 144–45.

CHAPTER 8

1. There are many alternative translations of the oath. The quotations in the text are from the version in John Chadwick and W. N. Mann, *The Medical Works of Hippocrates* (Oxford: Blackwell, 1950), p. 9.
2. George E. Hopkins, *The Airline Pilots: A Study in Elite Unionization* (Cambridge: Harvard University Press, 1971), p. 1.
3. Milton Friedman, "Some Comments on the Significance of Labor Unions for Economic Policy," in David McCord Wright, ed., *The Impact of the Union* (New York: Harcourt Brace, 1951), pp. 204–34. A similar estimate was reached more than a decade later on the basis of a far more detailed and extensive study by H. G. Lewis, *Unionism and Relative Wages in the United States* (Chicago: University of Chicago Press, 1963), p. 5.
4. Hopkins, *The Airline Pilots,* p. 2.
5. John P. Gould, *Davis-Bacon Act,* Special Analysis No. 15 (Washington, D.C.: American Enterprise Institute, November 1971), p. 10.
6. Ibid., pp. 1, 5.
7. See Yale Brozen and Milton Friedman, *The Minimum Wage Rate* (Washington, D.C.: The Free Society Association, April 1966); Finis Welch, *Minimum Wages: Issues and Evidence* (Washington, D.C.: American Enterprise Institute, 1978); and *Economic Report of the President,* January 1979, p. 218.
8. See Milton Friedman and Simon Kuznets, *Income from Independent Professional Practice* (New York: National Bureau of Economic Research, 1945), pp. 8–21.
9. Michael Pertschuk, "Needs and Incomes," *Regulation,* March/April 1979.
10. William Taylor, Executive Vice-President of the Valley Camp Coal Company, as quoted in Melvyn Dubofsky and Warren Van Tine, *John L. Lewis: A Biography* (New York: Quadrangle/New York Times Book Co., 1977), p. 377.
11. Karen Elliott House, "Balky Bureaus: Civil Service Rule Book May Bury Carter's Bid to Achieve Efficiency," *Wall Street Journal,* September 26, 1977, p. 1, col. 1.

CHAPTER 9

1. John Stuart Mill, *Principles of Political Economy,* vol. II, p. 9 (Book III, chap. VII).
2. Andrew White, *Money and Banking* (Boston: Ginn & Co., 1896), pp. 4 and 6.
3. Robert Chalmers, *A History of Currency in the British Colonies* (London: Printed for H. M. Stationery Office by Eyre & Spottiswoode, 1893), p. 6 fn., quoting from a still earlier publication.
4. A. Hinston Quiggin, *A Survey of Primitive Money* (London: Methuen, 1949), p. 316.
5. White, *Money and Banking,* pp. 9–10.
6. C. P. Nettels, *The Money Supply of the American Colonies before 1720* (Madison: University of Wisconsin, 1934), p. 213.
7. White, *Money and Banking,* p. 10.
8. Paul Einzig, *Primitive Money,* 2d ed., rev. and enl. (Oxford and New York: Pergamon Press, 1966), p. 281.
9. See Chapter 2.
10. See Phillip Cagan, "The Monetary Dynamics of Hyperinflation," in Milton Friedman, ed., *Studies in the Quantity Theory of Money* (Chicago: University of Chicago Press, 1956), p. 26.
11. Eugene M. Lerner, "Inflation in the Confederacy, 1861–65," in M. Friedman, *Studies in the Quantity Theory of Money,* p. 172.
12. Elgin Groseclose, *Money and Man* (New York: Frederick Ungar Publishing Co., 1961), p. 38.
13. John Maynard Keynes, *The Economic Consequences of the Peace* (New York: Harcourt, Brace & Howe, 1920), p. 236.
14. Robert L. Schuettinger and Eamon F. Butler, *Forty Centuries of Wage and Price Controls* (Washington, D.C.: Heritage Foundation, 1979).
15. The reason: a policy of trying to maintain a fixed exchange rate for the yen in terms of the dollar. There was upward pressure on the yen. To counter this pressure, the Japanese authorities bought dollars with newly created yen, which added to the money supply. In principle, they could have offset this addition to the money supply by other measures, but they did not do so.

CHAPTER 10

1. Raoul Berger, *Government by Judiciary* (Cambridge: Harvard University Press, 1977), pp. 1, 408.

2. *Lectures on the Relation between Law and Public Opinion* (1914 ed.), p. 302.

3. "Boom Industry," *Wall Street Journal*, June 12, 1979, p. 1, col. 5.

4. *Lectures on the Relation between Law and Public Opinion* (1914 ed.), pp. 257–58.

5. Milton Friedman, "Monumental Folly," *Newsweek*, June 25, 1973.

6. Friedman and Schwartz, *Monetary History*, p. 46.

INDEX